Become an ace at maths

A guide for pupils, students, parents and teachers

Dr Kobus Maree

J.L. van Schaik

Published by J.L. van Schaik (Pty) Ltd,
1064 Arcadia Street, Hatfield, Pretoria
All rights reserved
Copyright © 1992 Dr J.G. Maree

No part of this book may be reproduced, stored in a retrieval system, or transmitted, in any form, or by any means, electronic, mechanical, photocopying, recording or otherwise, without the prior permission of the publisher.

First edition 1992

ISBN 0 627 01839 4

Cover design by Barrett Joubert
Set in 10/12 pt Palatino by A1 Graphics, Pretoria
Printed and bound by Creda Press, Cape Town

WHY SUCH A BOOK?

Doctor Anton Rupert once said that he who does not believe in miracles, cannot be a realist. I am of the opinion that his words have a special meaning for pupils, students, their parents and teachers of mathematics.

One often hears that someone "is mathematically gifted and that is why he excels at it" or "he simply does not have a gift for maths and consequently cannot do well at it". Nothing could be further from the truth. Actually, there are two types of students and pupils: those who WORK HARD in mathematics and achieve commensurate results and those who do not work at the subject and as a result do not achieve good results. In this book I shall try to prove this statement.

When the brilliant mathematician Poincaré was at the peak of his career, he did the Binet Intelligence Test – and fared so badly that he was dubbed an idiot. Of David Hilbert, one of the greatest mathematicians of all times, one is told that his colleagues were always amazed that he took so long to understand new mathematics. In order to understand something, he had to work through it right from the start every time. Therefore, take courage!

Those who expect to find another series of examples of mathematical problems in this book, are going to be disappointed. I believe that there are more than enough textbooks which satisfy this need.

In this book I shall rather try to give pupils fresh courage. I shall try to illustrate why I am convinced that you and I do have the ability to do well – in other words, demonstrate how the ordinary pupil or student can become an achiever in mathematics.

What do I mean by the word achiever? I mean that you can take mathematics *up to the level you need to take it* – that you can take as much mathematics as you will need to follow the career of your choice. Naturally, we cannot all become professors in mathematics – but then very few of us want to.

Therefore, the theme of this book is: Help is available, you have a great moral responsibility to take mathematics up to the level that you can. Moral responsibility? Yes, in the sense that you owe it to yourself to do everything in your power to realize your God-given potential – in service to mankind.

Lastly, I would like to introduce you to Annie. Today she is a first-class medical student. Annie's teacher advised her to drop mathematics in standard seven – Annie did badly at it and "hated the subject". After a month in standard eight (without mathematics) Annie came to see me – she just wanted a last opinion on the matter. To cut a long story short: Annie passed matric three years later with a distinction in mathematics (higher grade). In this book I shall try to pass on our winning recipe in a simple way.

> "Truly it is not knowing but learning,
> not possessing but acquiring,
> not being there but getting there,
> which gives the greatest delight."
>
> Gauss

CONTENTS

1 To the parents of a child taking mathematics — 1

2 The parent's role in his child's achievement
in mathematics — 9

3 What are the main causes of problems in the
mathematics class? — 17

4 Motivation in mathematics — 35

5 Can I take mathematics? Should I take it? — 44

6 Mathematics anxiety, concentration and humour — 55

7 Mathematics, language, self-confidence and music — 62

8 Guidelines for becoming an ace at maths — 74

9 The part a mathematics teacher can play in making
his pupils achievers in mathematics — 94

10 Mathematics and culture. Are mathematics problems
learning problems or achievement problems? — 106

Addenda
A : List of examples of principles, rules and demonstrations
which you must always know by heart — 114
B_1: Mathematical symbols — 115
B_2: Mathematical terms — 116
C : Extract from model example file — 117
D : Errors file — 118
E : Fractions — 123

Bibliography — 126

CHAPTER ONE

To the parents of a child taking mathematics

Introduction

Do you recognize the following set of marks?

English	HG	75%
Afrikaans	HG	70%
Biology	HG	67%
Science	HG	64%
Accountancy	HG	71%
Mathematics	HG	28%

These were Peter's[1] marks when he came for extra help in mathematics earlier this year. However, they could just as well have been those of your child – more or less. Believe me, I would like to welcome you to one of the biggest clubs in the world – the club for parents whose children have problems with mathematics.

During the years that I have taught mathematics (and I still do), as well as the years that I have been involved with the Career Development Centre at the University of Pretoria, the following remarks and questions by parents have virtually become part of my daily communication:

- Mathematics is a nightmare for my child.
- My child really struggles with mathematics.
- Where can I find extra help in mathematics for my child who is struggling?
- How can I help my child do better in mathematics?
- Why is my child struggling with mathematics?
- Everybody says my child should pass mathematics, but he simply cannot do it; what is the matter?

1. Pseudonyms are used throughout although the cases are based on fact.

- Does my child have the ability to pass mathematics?
- How can I make my child believe that he can do mathematics?
- How does a child develop self-confidence in mathematics?
- My child's teacher refuses to give him extra attention or to explain something again; he keeps saying that there is no time. What am I to do?
- My child now wants to take mathematics at standard grade. Should I allow this?
- How do I motivate my child to work harder at mathematics?
- My child simply dreads mathematics.

I can go on indefinitely. The fact of the matter is that it is almost every parent's fondest wish that his child should do better in mathematics. Every right-minded parent knows how important this subject is to his child – hence the concern.

This book approaches the subject from a unique angle. For the first time, the struggling child and his parent are given a handy guide and the role of the teacher is also spelled out clearly. The psychology behind mathematics, the main reasons why children fail as well as the means to put them back on the winning path, are discussed against the background of in-depth research. The writer and many teachers[2] know the feeling of satisfaction when a pupil starts to do well again, but they also know the feeling of despondency that is inevitable when a pupil gets bogged down in a situation from which he can rarely escape *by himself.*

What is more, it is my deepest conviction that most children can pass mathematics, contrary to the popular and convenient excuse that the subject is only meant for a small group of elite pupils.

I shall now explain briefly why it is not only or necessarily the "clever"child (or the child with a high IQ) who can do mathematics.

Intelligence and mathematics

When a child does the individual (and most reliable) intelligence test, he[3] in fact does nine sub-tests. The fourth test comprises *problems.*

---

2. This term refers to any member of the teaching profession.
3. When referring to a child in this book, I shall always use "he" and "his". This is only done for the sake of brevity and one can read "she" or "her".

Your child might do badly in this test and then you are likely to hear: "Do not expect too much from your child, he is probably not capable of doing well in mathematics". However, this is a highly controversial statement which should not be accepted unconditionally.

The following table shows the results achieved by a number of children, in the standards indicated, in the mathematics part of the IQ test. One would obviously expect that a pupil who obtains high marks would accordingly do well in mathematics. When comparing these pupils' marks in mathematics with their "potential" (as indicated by the marks obtained in the IQ test), one finds, for example, that Gerda obtained 16 out of a possible 17 in the IQ test, but that she performs poorly in mathematics. One would hardly expect Gerald, with 11 out of 18, to obtain a brilliant 87%! Henry is exceptional: he scored 8 marks out of 17 in the "maths" section of the IQ test but scored 81% in the higher grade!

Name	Standard	Test score (problems)	Marks in mathematics
Peter	7	10 out of 19	81% (SG – standard grade)
Gerda	9	16 out of 17	35% (HG – higher grade)
Gerald	8	11 out of 18	87% (HG)
Marina	10	11 out of 17	76% (HG)
Henry	10	8 out of 17	81% (HG)

What does this table demonstrate? That the score a child obtains in the IQ test does not predict how that child is going to perform in mathematics. Of course, there often is a similarity between the IQ score and the maths mark – but this is certainly not always the case. The IQ score does not correctly predict the child's future performance in mathematics. Why not?

Why do some children perform better and others more poorly than their parents and teachers expect them to?

Tests which measure "mathematical ability" or "mathematical aptitude" are designed primarily to measure a child's ability to

3

think abstractly in terms of figures. The calculations he is asked to do are intended to give an indication of intellectual development, specifically mathematical development.

In order to work with figures, a child's mind has to be clear. He must be able to concentrate in spite of the thousands of other things which might distract his attention. *When these concepts, sums, calculations or sketches become more difficult and when the meaning of the material is unclear or complicated, the ability to concentrate is of vital importance.* On page 60 you can read more about the difference between concentration and attention; at this stage suffice it to say that there is an important difference between the two concepts.

It is especially important to know in what way insufficient concentration is likely to affect your child's performance in mathematics.

Concentration and performance in mathematics

When I am testing a child's intelligence and I reach the "mathematics sub-test" I so often hear the following remarks (I must add, when concentration during the test has not been good):

- I just cannot get on in maths.
- I am the weakest at maths in my class.
- I do not know anything about maths.
- Sums have always been a problem to me.
- I am stupid at maths.
- I'll never be able to do maths.
- I hate maths.

This child is hiding behind a mask of indifference. He will deliberately try to look bored and indifferent, and he will show no interest in mathematics. Yes, your child's concentration has been adversely affected by poor instruction, a lack of success and other factors which will be discussed further on.

Good performance in this test and in mathematics at school is very often the result of good concentration. A calm, collected child invariably performs better than a child who shows signs of anxiety. A low score in this test often indicates an inability to concentrate or poor focusing of attention.

Obviously, anxiety about mathematics can originate from domestic problems (and sometimes from your child's rebellion against your authority or even authority in general). If parents quarrel continually (with one another or with the child) or even have a fight the evening before the test, it can have a seriously detrimental effect on the child's performance. As a matter of fact, even transient, short-lived emotional disturbances can have a seriously detrimental effect on your child's performance in the mathematics part of the IQ test and at school.

When your child fears his teacher and his moods, when he is denigrated before the class, humiliated in front of his friends and gets the impression that he is "stupid", if his experience is that very little is expected of him, if he is intimidated by a teacher who wants to show the pupils how good he really is at mathematics, if he gets the impression that the mathematics examples used in class by the teacher exclude him, if he has a poor relationship with his mathematics teacher, or if he does not dare to ask questions in class, then his problems with concentration are practically a logical consequence of the situation.

But I would like to return to another factor, which concerns you as the parent, one which has a direct influence on your child's performance in mathematics: parental expectations.

Parental expectations and performance in mathematics

Do you recognize one or more of the following attitudes which parents often adopt:

- I could not (did not want to) do mathematics and I know the problems one faces without mathematics, therefore my child must take maths.
- I do not care what others say, my child will take mathematics.
- A boy cannot do anything without mathematics.
- It is not very important that my daughter should take mathematics.
- My poor child works himself to a standstill but he still does not do well in mathematics.
- My child has the worst mathematics teacher in town.
- The teacher never even looks at my child's homework books.
- My child should be able to work on his own in mathematics, otherwise he will not be able to work without help when he is at university or technikon: I am not going to rob him of his ability to work on his own.

- It is a fact that a child can either do mathematics or cannot, there is nothing to be done about it – my child takes after me, I could never do mathematics either.
- Mathematics is not something you can learn – you either understand it or you do not.
- My child can do whatever pleases him – if he does not want to take the subject, it is his problem.

And so I can continue. The fact of the matter is that each of these statements, like all good propaganda, contains a germ of truth. But this does not make them true or reliable!

Let us deal with a few of the most popular misconceptions concerning achievement in mathematics.

1. "My child has to learn to work on his own, otherwise he will not pass at university in any case. Moreover, the children's homework is my wife's department."

   Correction

   As the father, you play an important role in the mathematical development of your son or daughter. And I do not mean incidental or chance interest, now and then, but sustained attention and interest throughout the child's school career and further studies. Research shows time and again that boys achieve better results in mathematics when their fathers share this interest with them. Children do not learn spontaneously – they must be taught to learn, to be industrious.

2. "I did not take advantage of my opportunities in mathematics and today I regret it; my child *will* take the subject and perform well" or, on the contrary, "My child takes after me. I could never do mathematics either – how can I expect him to take mathematics?"

   Correction

   There is *no* reason why a child *should* do well if his parents did not perform well in mathematics. Nor is there any reason whatsoever why a child should not perform well if his parents did not do well in the subject.

3. "Without mathematics a child cannot do *anything*."

   Correction

   There are many careers a child can follow without passing matriculation mathematics – ask any vocational guidance counsellor.

4. "Why all the fuss about mathematics? – it is a school subject like any other."

Correction

On the other hand, mathematics does of course open many doors to a child, doors which would otherwise stay shut.

5. "Only the very clever ones have the chance to take mathematics up to matric."

Correction

Most children can pass matriculation mathematics – and an infinitely greater percentage than at present can pass the subject in higher grade.

6. "It is one of those things: one child has a knack for mathematics and *can* do it, the next does not have the knack and will never be able to perform well in it."

Correction

There is no such thing as a pupil who can do mathematics and one who cannot do it. It is just not true. On the other hand, there are pupils *who work hard and pass and other pupils who work less hard and do not pass.*

7. "You either have it or you do not – self-confidence has nothing to do with performance in mathematics."

Correction

External factors such as motivation and your child's emotional condition, his attitude towards school work and authority, his self-image and his self-confidence can have *a seriously adverse effect on his performance* if something is wrong there.

8. "Girls are just weaker at mathematics than boys."

Correction

There is no reason why girls should necessarily perform less well than boys in mathematics.

9. "The teacher is paid to teach my child mathematics – if the child does not do well, the fault lies with him."
"The whole class does very poorly in mathematics at any rate – the problem lies exclusively with the teacher."

Correction

It is certainly not the sole responsibility of the child's teacher to lead him to success in mathematics. Activities like signing his homework books, making sure that he makes corrections, signing of *all* tests (including the smallest class tests), attending parent evenings, following up problems *with the teacher himself* and checking your child's study times, are also your responsibility.

And talking of the teacher: Please remember at all times that one can always find pupils who do perform well in your child's class (though he may be under-achieving/doing poorly in mathematics) despite possibly serious shortcomings in the instruction.

10. "I hated mathematics, I am unhappy in my work and I often tell my child – that will teach him to follow a different career."

Correction

Your attitude towards work in general and mathematics in particular will probably rub off on your child – sufficient reason for you to speak of these matters in his presence in a balanced way.

To presume for convenience's sake that many pupils are "stupid", not capable of doing mathematics, that they cannot understand the subject and therefore should rather not take mathematics, is a dangerous escape. It is generally not very helpful to try and treat pupils and their problems with mathematics in terms of low intelligence, low mathematical aptitude, perceptual problems or just learning problems. Of course these factors play a role but – if we are really serious about mathematics and its teaching, and especially if we care about the child behind the subject – then it is of the greatest importance that we should always go to the root of the problem.

Conclusion

In this chapter I have focused your attention on the fact that, as his parent, you have a particularly meaningful part to play in your child's performance in mathematics. In the next chapter I am going to look more specifically at what you can and should do to create the circumstances for optimum performance in mathematics.

CHAPTER TWO

The parent's role in his child's achievement in mathematics

Introduction

You have probably already taken the most important step in handling your child's problems with mathematics. You have admitted that there is a problem and you now have to start *solving* it. By this I mean that you should never just *resign yourself to a problem*. Your child should never get the impression that he is subject to some kind of fate in mathematics. No, rather teach him that he should do something about it, as with any other problem in life.

My approach in this book is the same throughout. *Under no circumstances should you do your child's mathematics or try to do it for him*. In the majority of cases parents are simply not equipped to help pupils up to matric level. And it is potentially really extremely destructive – for reasons I shall discuss later. However, this certainly does not mean that you cannot help him achieve good results in various other ways – as you will see in this and other chapters. Whether you encourage him, whether you check his "corrections file" and his "model examples file" and congratulate him on his careful and neat work, whether you sign off his homework book or test and add some comment (if he gives permission for that comment!) there is much that you as parent can and should do in this regard. In any case, I want to ask you to read the whole book – it contains very little that you will not be able to understand, but much that will give you new insight.

First of all, let us look at what experts consider to be the most obvious course of action if you want your child to be an achiever.

Encourage your child

After doing many years of research on this subject, Dinkmeyer and Dreikurs say that every parent should encourage his child in the first place, and especially in the following ways:
- Value your child for what he *is*.
- Show faith and trust in your child – this develops his self-confidence.
- Show your faith in your child's ability; try to keep his confidence and build his self-respect.
- Give him the necessary recognition for any achievement in any respect, especially achievement in mathematics.
- Give your child credit for his efforts.
- Use the family set-up to facilitate and to promote your child's development.
- Integrate the family in such a way that the child knows *exactly* where he fits in.

Who taught you that one can subtract 15 from 9?

- Help your child develop his abilities – but set a pace suited to his age group: physically, mentally, psychologically and normatively. For example, you can stimulate your toddler's mathematical development by letting him complete jig-saw puzzles and build with blocks – and by ensuring that you use the right names throughout, for example corner blocks, side blocks and middle blocks.
- Give your child ample recognition for his strong points and for his potential.
- Weave your child's interests into instructions that you give him.

I especially want you to take note of the fact that *nowhere, but nowhere, do I require that you as parent need to have a particular qualification in mathematics in order to turn your child into an achiever in mathematics* ...

But it is of equally great importance that your child should feel safe at all times – a lot of cold figures are enough to completely strip any otherwise stable child of his feeling of safety!

Help your child to feel safe

The authors Cruikshank, Fitzgerald and Jensen stress the fact that parents and teachers often feel that it is their duty to point out children's mistakes at all times. However, they believe that it is much more important and more constructive to look for children's strong points and assets. Especially where mathematics is concerned, it is better to teach children the correct concepts without damaging their delicate egos with excessive condemnation of their mistakes.

I strongly recommend the following guidelines to prevent failure and the feeling of I-am-a-failure, and to help children become self-confident learners:
- Accept your own mistakes and faults, admit to them in front of your children. Always point out to your child that failure can be of great help in the process of learning, but is unimportant in measuring his value as a person.
- Make allowances for failure and handle it as a natural part of the learning process. Allow your child to make mistakes and to learn from them, but avoid making him feel embarrassed about a mistake.

- Create opportunities for children to be successful. By achieving success, your child eventually develops so much self-confidence that he retains his self-confidence and feeling of safety even at those times when he does not pass or succeed.
- Lay much less emphasis on the importance of doing everything "right". Do not keep blaming your child if he systematically does not come up to your expectations. *Rather expect him to perform well within the context of his abilities.*
- Accept that every child is an individual with a right to respect. Believe that every child deserves respect. Listen to your child if he wants to share his ideas and feelings with you and react warmly and sincerely.
- Create opportunities for your child to make decisions regarding his own life. Make sure that these decisions are sincere and that your child is prepared to accept the consequences of his decisions *himself*.
- Discuss failures both individually and in the family context. Always try to reach an agreement on the way in which failure should be handled and honour these agreements.

I cannot stress the importance of these measures enough. I often address groups of people, and I always add the following remarks in conclusion:

It should be stressed that personality traits or characteristics, like the will to perform well in mathematics, perseverance, self-discipline, decisiveness, a sense of duty, determination and temperament are of cardinal importance and will eventually be co-responsible in determining whether he achieves success or not. Aptitude/intelligence/ interest in themselves are no guarantee for success.

Choosing to study, and whether or not to study hard, is your child's responsibility. He will eventually have to choose and do the work himself (with the help and support of his parents and other significant persons). No person can or may make the important decisions in life on someone else's behalf.

In mathematics, success sometimes hinges on small things too – and one of them is *character*.

At this stage I want to draw your attention once more to the fact that I never demand that you should do the child's mathematics for him or even with him. In other words, leave it to those who are qualified to do it. You can do the most important things – the rest will follow automatically.

But I want to return once more to your child's achievement in mathematics and his self-confidence.

Help your child to develop self-confidence

What does a child with self-confidence look like? According to Van Rooyen & Schultze a child with self-confidence

- believes in himself as a person, in his capabilities and convictions;
- asserts himself but gives others a chance too;
- is more independent and creative;
- is convinced of his own opinions and conclusions;
- participates rather than listens passively in a group and feels free to air his views – even if he does not agree with the others;
- does what he thinks is right when he holds a different opinion;
- does listen to the ideas of others – he does not feel threatened;
- makes friends easily;
- is not engrossed in his own personal problems;
- dares to do new and difficult things – physically and psychologically (in other words, on the playing field and in the classroom);
- is motivated to win, but is prepared to lose and to learn something from it;
- is not overly self-conscious.

This self-confidence is assured by parents who

- have a well-developed self-image;
- have enough self-confidence;
- are emotionally stable;
- are at ease in handling their children;
- have a positive attitude towards their children;
- adapt to changing circumstances with ease;
- are encouraging and supportive of their children;
- encourage autonomy and *emotional independence* in their children;
- have definite rules at home, teach their children to obey the rules – and do so themselves;
- give the appropriate recognition to the rights and opinions of their children;
- always practise the following rule in connection with discipline: "The punishment should fit the offence."

How does all of this tie in with mathematics?

To start off: Mathematics is the one subject where one can easily identify a child with emotional problems. Pupils without self-confidence and with a low self-image often struggle with mathematics. I recently experienced a striking illustration. After therapy the self-image of a boy improved drastically, and his maths marks improved from 35% to 88% within eight months.

There is a particularly strong link between a child's self-image and his performance in mathematics. If he does badly at school – even if it is as a result of laziness – he soon believes that he is stupid and develops a sense of inferiority.

Together with this he usually has a lower level of aspiration and achievement. It can be summarized as follows:

- The child who learns to work at home, learns to believe in himself, develops a better self-image and also stands a chance of working harder at school.
- The child with a better self-image who works harder at school, will achieve better results.
- The child who performs better develops higher aspirations, gains entrance to more exclusive fields of study and career orientation and stands a much better chance of achieving happiness in life.

You have probably noticed that I make a strong connection between self-confidence, achievement and – yes, of course, hard work.

Hard work and achievement in mathematics

At the outset: Success – even for the most gifted person – never comes suddenly; it is the result of many hours of hard work. The truth is simply that laziness is not inherited but is learned.

Parents are often blamed unfairly for many of their children's problems – but in the case of a child who is too lazy to do mathematics, it really is the fault of the parent. Why is a child lazy to do his mathematics? Because children, especially in the present-day affluent society, are never really taught to exert themselves. At primary school level very few demands are usually made of the child. Especially for the highly gifted child the standard of work is sometimes much too low. Furthermore, the amount of homework they are given seldom demands much exertion on their part – and then mother is often at hand to do the homework for the child or to help him with it. That can be the start of a vicious circle. An intelligent child who does not learn to work, in fact becomes lazy – lazy to learn and lazy to work. The lazy pupil too often becomes a lazy student and, later still, a lazy adult who never realizes his full potential.

Mathematics is an extremely abstract subject, and requires your child to work with figures and symbols. To perform well, he must practise regularly. To put it plainly: The more your child works, the better he performs.

This chapter would be incomplete if I did not also examine the place of the girl in the mathematics class thoroughly.

Your daughter and mathematics

Visser stresses the fact that in many subtle ways (and sometimes even less subtle ways!) girls are discouraged from taking mathematics and from doing well in it. Because mathematics is largely regarded as a male field, it follows that girls would rather try to prove themselves in other fields. Girls are inclined to take part in activities which raise their self-esteem – and this often means that girls who experience problems with maths will rather take another subject that is more likely to boost their self-confidence.

In an authoritative research report Visser reaches the following conclusions that are well worth noting:

- Parents have higher expectations of their sons' performance in mathematics than of their daughters'.
- Mathematics is considered a male activity.
- Parents' expectations and encouragement are an important indicator of performance in mathematics, especially in the case of standard 7 girls who are on the point of making a choice about mathematics as a matric subject.

The results of these misconceptions are far-reaching – just think of those careers in which the largest shortages exist and what a significant increase in female professionals in these fields would mean to the country.

There are numerous reasons why women are discriminated against in mathematics careers. Unfortunately there is not enough time or space to discuss these reasons in depth. However, I do want to express the opinion that it is more than high time that something is done about this sad state of affairs.

There is no reason why you should expect less of your daughter than of your son in the mathematics class. On the contrary, I want to urge you to have the same expectations of boys and girls if they are of the same age and have comparable aptitude and ability. Otherwise you are simply acting unjustly towards your daughter.

Finally: What about extra lessons in mathematics?

Extra classes in mathematics

Firstly: Do not hesitate to seek help if your child starts to experience problems in mathematics. However, this search should start with the child's own teacher. First try to set the matter right because the teacher is the person who should best be able to understand your child's particular problems. He will be able to tell which parts of the work the child does not understand and whether he is perhaps simply lazy.

Secondly: Extra lessons do not replace the teacher and certainly do not mean that your child can now stop working.

Teachers usually have no objection to a pupil receiving extra lessons in mathematics. But you have to realize that extra lessons are not like a single visit to the doctor – the child will have to attend these lessons over a period of time and at fixed times. The very best person for such lessons is somebody who has a thorough knowledge of mathematics, pedagogics and psychology. This enables him to determine whether the cause of the problem is perhaps something like a weak self-image. Please read pages 68 and 69 on what I have to say to your child in this regard.

Conclusion

There are various measures every parent can take to ensure that his child has the best chance of performing well in mathematics – measures which lie within the reach of every parent.

In the next few chapters I am going to talk to your child, but I would like you to read these chapters with the same attention and concentration as the child. It is just possible that I can convince you that your child is indeed a potential maths whiz!

But Dinkmeyer and Dreikurs have the last word: "Mathematical training of the child can be effective in a democratic setting. Simply stated, it must include a respect for order, avoidance of conflict and encouragement."

CHAPTER THREE

What are the main causes of problems in the mathematics class?

Introduction

Before you read any further, remember that each of us has his own, personal style of doing things. One likes this pop star, the other that one and a third yet another. Your taste in clothes, your choice of a girlfriend or your eating habits are all strictly private matters. This principle is also valid in mathematics. Each of us has a personal style of doing mathematics – and especially of making mistakes, of handling problems. It is of the greatest importance that you should find out what the flaws in your style are – and remember throughout that it will take *time, time* and *practice, practice, practice* to eliminate these spoilers.

Why do you make mistakes in mathematics?

Of course there can be many reasons for mathematical errors:

- Perhaps you simply do not understand the work. In this case it is imperative to remember that you should try to tackle the problem *immediately*. An apparently small problem can easily grow into an enormous one. Each day's work is built on the foundation of that of the previous day, each week's work is built on that of the previous week, the work of each month and each year is built on the foundation of that of the previous month and year, and it is sometimes extremely difficult to rid yourself of basic errors in a later stage of your school or subsequent career.
- You may learn to make certain mistakes. You very often learn certain habits, like doing your work over-hastily, untidily,

17

carelessly or inaccurately; drawing lines and sketches free-hand; erroneously marking incorrect work as correct; not doing corrections properly; and many others on which I shall dwell later on. These habits have to be broken again, but be forewarned: This can be an extremely laborious process.

- Perhaps you have never really adapted properly to your present school. This can lead to other problems, like insufficient discipline or a negative attitude of "I do not like school in general and mathematics in particular".
- You may have a problem with your eyes or ears, you may be listless, may follow an unbalanced diet, may often be absent (and therefore miss out on essentials in the mathematics class). In this case I would suggest that your parents take you to a medical doctor, if only to make sure that there is no medical reason for your problems.
- Your background (family and otherwise) may be your most limiting factor. Do be very honest and ask yourself the following questions:
  - Is my language development limited? One usually knows very well when this is the case. And the solution to this kind of problem lies partly in *reading* as much as possible, as often as possible, in listening to good radio programmes (with *attention* and *concentration*) and in ensuring that you can follow the dialogue when watching television.
  - Do I come from a broken home or unstable family? If so, it is desirable that you ask your school guidance counsellor or minister of religion, or any other adult whom you trust, to help you.
  - Is my attitude towards school generally negative? This does not have to be the case, since every school provides for particular needs, yours too. Ask the help of a sympathetic teacher in this regard.
  - Do my parents encourage me often enough? If not, I would suggest that you take up the matter with *them* and talk about it. You might even do it in writing – and remember that they are only human, they do not always know how you experience them as parents and the way they treat you.
  - Do I get enough sleep? Regardless of what you may think, a good night's sleep is important.
- Sometimes you may not even be aware of the fact that something is bothering you. But it could just be that at the

time that you were doing factorization with advanced trinomials, you were so upset by your parents' quarrelling that you lost track and are still struggling to find your way. Or you cannot concentrate but you can think of no earthly reason why not. In such a case, it may be desirable to consult a psychologist – he may just be able to determine what the unconscious block is.
- If there are things that bother you, and which you know about, then you must talk to somebody about them. In other words, if there is unnecessary stress and tension in your life (and you alone know exactly what causes the stress) then it will certainly influence your performance in mathematics. Mathematics, more than any other subject, is negatively influenced by stress, especially so because you are usually required to think abstractly if you want to achieve success in mathematics.

The bottom line should always be the following: *Do something. Never passively resign yourself to your lot.*

Let us discuss briefly some of the more common mistakes in classroom mathematics. *Please! Get yourself a file right now (see addendum D at the back of the book) and make a note of your particular errors when you recognize them.*

First let us see how well you know yourself. Quickly write down the ten most common errors (or combinations of errors) you are inclined to make when doing mathematics:

1. _____

2. _____

3. _____

4. _____

5. _____

6. _____

7. _____

8. _____

9. _____

10. _____

At the end of the chapter you can compare this list with the one you compile then.

It is now time for us to look at more specific errors all pupils make from time to time to a greater or lesser extent.

Examples of standard errors and other reasons for poor performance in mathematics

To start with: You and I are human, imperfect and fallible, limited as regards life and thought. That is why we make mistakes – and the Creator knows that, that is why He gave us the choice to correct these mistakes or to let them go by.

At all times be prepared to *discuss* your mistakes with your teacher, to tell him exactly why you do something, what you have in mind with a specific step, what your line of thought is, what you had in mind when you made the mistake. No mistake should *ever* be the subject of ridicule, no mistake should ever make you feel bad or inferior (stupid) – mistakes are your way of making sense of a sum, of finding a meaningful solution to a problem. *Talk about it!*

And by the way:

> *Guard against becoming too accustomed to the letters x and y,* the notations $f(x)$ and $g(x)$ and other overly familiar and stereotyped usages. It sometimes causes you to become unnecessarily confused with notations such as m, n, $t(x)$, $q(x)$ and $\phi(x)$.

To begin with and without wanting to confuse you with technical detail, I would like to highlight those mistakes which are made sporadically as a result of negligence, undue haste or common carelessness. Let's call them slips. Slips are often made by experienced as well as inexperienced mathematicians. We often correct them spontaneously and trace them very easily (Olivier).

Then there are mistakes which are *symptoms* of underlying misconceptions[1]; they are often due to erroneous planning, and occur systematically in the sense that similar circumstances often lead to the same mistakes. Misconceptions about the underlying principles of mathematics are often the *cause* of mistakes.

The value of mistakes is that they allow us to identify the underlying misconceptions, if we only take the trouble, after which we can make a conscious effort to rid ourselves of these misconceptions.

Look at the following example:

Solve for x: $2(3x - 1) - 5(x - 2) = 4x - 4$

(If you make a mistake in copying down the sum, then I regard it as a slip!)

Look at the correct solution:

$$6x - 2 - 5x + 10 = 4x - 4$$
$$\therefore \quad 6x - 5x - 4x = -4 + 2 - 10$$
$$\therefore \quad -3x = -12$$
$$\therefore \quad x = 4$$

---

1. Here is an interesting error to illustrate this fact: In your first paper you get 90/200 and in your second 150/200; i.e. 240/400. But if you want to add 1/2 to 1/2, then the answer is not 2/4 but 1!

I shall discuss some mistakes which may occur:

1. $6x - 2 - 5x - 10$ (slip, or you do not realize
   $= 4x - 4$ that: $- \times - = +$)

2. $-3x = -12$ (basic insight mistake:
   $\therefore \quad x = -12 + 3 = -9$ $-3x \neq -3 + x$)

3. $6x - 1 - 10x - 2$ (basic mistake: you do not know
   $= 4x - 4$ that: $a(b + c) = ab + ac$)

4. $-3x = -12x$ (slip!)
   $\therefore \quad x = 3$

5. $2 + 3x - 1 - 5 + x - 2$ (basic mistake: you do not know
   $= 4x - 4$ that: $a(b + c) = ab + ac$)

Of course you can make many other kinds of mistakes; we are presently going to discuss a few. However, I want to urge you *never to allow your teacher to take his pen and do the "wrong" sum for you.* Your examiner is never interested in discovering how good your mathematics teacher is or was at maths, only how good or bad you are at it!

One of the most important mistakes in mathematics is that a pupil does not know the basic principles.[2]

Insufficient knowledge

Whether the cause lies in inadequate teaching methods, not learning enough, environmental factors or whatever, I regard insufficient knowledge as the first mistake. In order to proceed to the solution of problems, you simply have to give your brain a reasonable chance (read more about this on page 80) to gain insight into the necessary concepts. This means that, at all times, you simply have to abide by the following principles:

- Know the language and symbols of mathematics. If you do not know exactly what the symbols Σ or $\frac{dy}{dx}$ mean, you cannot even try to work with them. If you are not sure what the words median, diagonal, attitude, bisector and perpendicular bisector mean, then you cannot possibly try to do geometry.

2. Note: I do not even try to claim that this list of mistakes is at all exhaustive. I rather want to make you aware of the fact that *mistakes should always lead to greater insight.*

- Know certain specific principles. I mean principles such as $a^m \times a^n = a^{m+n}$, and $(ab)^n = a^n b^n$, in other words, laws, definitions and so forth, without the demonstrations.
- Know the prescribed demonstrations of your principles. If you do not know this, surely only one person is to blame.
- Techniques and skills have to be practised until they become automatic. I am referring to anything in class which you have learnt to do with ease and accuracy. Whether it is the application of a formula or handling your pocket calculator, you have to be able to do it *well*.

Have you ever seen an experienced typist at work? She will laugh if you ask her whether she sometimes searches for the keys. By sufficient practice the above principles simply have to become a part of you, in which case we talk of *autonomous processes*. *Whenever possible*, try to reach this level in mastering new mathematical principles, and half the battle is won.

> *There can be no passengers in the mathematics class – you can never "go along for the ride!"*
> *Success at maths begins with honest, hard work.*

The following kind of mistake was my personal "favourite" – I repeatedly made careless errors.

Careless errors

One is careless when one makes mistakes which cannot be ascribed to a lack of insight.

Why are you sometimes careless? There are various reasons for this:
- Your attention is distracted.
- You are bored.
- You do not complete your thought process. This occurs when you make an error in copying down a sum, when you transfer something incorrectly (especially at the end of an expression or an equation). Have you ever experienced the frustration of finding that your sum is perfectly correct – but that you did the wrong sum; or that you did not do a sum (while you could do it); or even that you made a careless calculation mistake in the second line of a long sum?

- The most important reason for careless mistakes is, to my mind, the fact that one does not concentrate fully when doing mathematics homework – I repeatedly find that careless pupils learn this style when they do their homework with the radio blaring or while chatting gaily.

The following three mistakes always make me smile when I come across them: The pupil believes against all odds that he has copied down the sum incorrectly, that there is a typing error on the paper (or in the textbook) or that the sum "cannot be solved".

Change of instruction

This simply means that you find the assignment too difficult, and then replace it with another, which you think is right (and usually much simpler!) A good example of this is the chap who replaces $(x - y)(y - x)$ with $(x - y)(x - y)$ because this is a more familiar type of simplification.

Regression

Regression usually takes place when the work becomes more difficult. You lose control of the situation and your line of thought becomes superficial, a kind of automatic way of thinking. For example, you learn that $0,4 \times 0,2 = 0,08$ but you do not know exactly why. This often leads to a situation where you make your own rules. Then, when you cannot remember what you have learnt, you make your own rules. I must add, this often means that you take the easiest (and not necessarily the most correct!) way out. While you are concentrating on one process, your attention lapses and negligent mistakes are made. Even concepts which you initially understood but imperfectly or superficially, can eventually lead to confusion and poor performance. That is one of the main reasons for making the following mistake: $0,4 + 0,5 = 0,09$.

Forgetfulness

Of course, forgetfulness is one of the most common reasons for the mistakes that pupils make.

I wonder whether there is one pupil who hasn't had "the answer on the tip of his tongue" but just could not remember it[3], has felt the anguish and frustration of sitting in an examination room only to realize that he just cannot remember the information he learnt and now needs to solve the problem. The golden rule in this case is: Revise the work you have done every day, sort out problems and make sure that you have a thorough knowledge of the concept overlearning. Read more about this on page 84. You often forget simply because you have not absorbed the concepts sufficiently.

In this book I repeatedly refer to the detrimental effects of emotional problems. But I do want to point out once again that your mathematical memory is the first to go if you experience an emotional upset. It is much more difficult to remember anything, but especially figures, formulas and mathematical concepts when something is bothering you. That is reason enough to try and handle any emotional problem *immediately* it occurs. And if you cannot do it immediately, make yourself a note in a conspicuous place and tell yourself: I am going to handle this problem on that day, I will not allow it to bother me now.

After-effect and anticipation

Look at the following two examples very carefully:

1. Factorize: $m^2 + 5m + 4 = 0$

   $\therefore (m + 1)(m + 5) = 0$. We call this *after-effect* – the 5 sticks in your mind and prevents you from going on to the next step in the solution process.

2. Find m: $m(m - 4) - 6(3 + 7m) = 9$

   $4m^2 - 4m - 42 - 42m = 9$. This is a case of *anticipation* – and can be ascribed to the fact that you read considerably faster than you write.

Inability to generalize (or generalize correctly)

This problem sometimes starts right at the beginning of algebra when your teacher leaves you under the false impression that $4x + 5x$ are simply abstract *letters*, that $2a + 3b$ is the sum of

---

3. Handle this problem in the following way: Try to think of something else and then return to the elusive answer. Otherwise, go through the whole alphabet – the letter B might help you to remember the name Bush.

2 apples and 3 pears. You must realize from the start that we are not making a fruit salad and that *letters are numerical variables* with which we can do algebra. In other words, figures become letters for the simple reason that we *can generalize with letters*, x can be any figure, 2 is ONLY 2!

Now look at the following examples:

1. $x^2 - y^2 = (x - y)(x + y)$

   If you cannot generalize and you are asked to factorize the following examples, then you really have problems:

   $x - y$

   $9x^2 - 16z^2$

   Or one that is even more difficult, like the following:

   $2(3 - m)^2 - 50(p + q)^6$ where you have to integrate quite a few principles to boot.

2. Primary school pupils learn at an early stage that at least two numbers are involved in an addition. This generalization interferes with later addition (read more about the phenomenon of interference on page 31) and gives rise to mistakes like the following:

$$
\begin{array}{cc}
453 & \quad\text{and}\quad & 236 \\
+21 & & +14 \\
\hline
674 & & 25 \\
\end{array}
$$

Can you see the reason for the wrong answers?

Do you see why it is of the *greatest* importance that you should *talk* to your teacher about your mathematics? Tell him how you go about getting the wrong answer.

In other words: Always try to reason back to the simplest, most general type of problem. Always try to give reasons for what you are doing, to find the general principle behind your specific sum. *It is my firm conviction that pupils who have mastered this ability, and who apply it at all times, will never have any cause to struggle with this subject again.*

In other words: *Talk* about your mathematics, chat about it, discuss it with as many people as possible. Give reasons for what you write, say it aloud to yourself, your parents, your friends, your teachers, your brothers and sisters. This brings me to the following very important mistake.

Careless and negligent language and writing

Read the pupil's explanation of the sum below:

Steps	Gerald's explanation
$$\dfrac{2(3x-2)}{3} - \dfrac{3x+7}{6}$$ $$= \dfrac{4(3x-2)-(3x+7)}{6}$$ $$= \dfrac{12x-8-3x-7}{6}$$ $$= \dfrac{9x-15}{6}$$	"Get the same figure under the line and multiply the figures above", instead of: *Find a common denominator and relate the fractions to this denominator – a fraction stays the same if we multiply or divide the numerator and the denominator with the same number.*
$$= \dfrac{3x-5}{2}$$	"3 can be divided into the numbers", instead of: *3 is a common factor to the numerator as well as the denominator* (LCM).

The fact is if you do not talk to your teacher or somebody else who understands the sum, nobody can tell what mistake you made, why you made it or how you can sort it out. To name something confidently in mathematics, is to understand it!

By the way, it is of the greatest importance that your mathematics teacher should *regularly* ask you to state *verbally* what you are doing, and to repeat *verbally* the steps he has done on the blackboard, from time to time. This is the most effective way to learn to "speak mathematics" – something which is of vital importance if you want to be an ace at mathematics.

Every term, word or symbol in mathematics has an important function and plays an important role. If you cannot interpret symbols, if you use them without being able to name them, you are guilty of VERBALISM. This means that you know what you can or should do with a symbol, but that you do not really understand what you are doing. The greatest danger in this respect is that you are missing out on the extremely important ability to generalize. In Chapter 7 you can read more about this, but remember that even simple words like "term", "expression", "equation", "numerator", "denominator", "least common multiple" and "largest common factor" are of cardinal importance in building your mathematics foundations. In other words, the use of language in mathematics is very specific – it is not open to different interpretations.

The next mistake is often found in pupils' geometry books.

Reasoning in circles

Look at the following example:

Demonstrate AB = DC and AD = BC

In △s ABC and CDA:

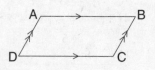

	Statement:	Reason:
1.	AB = DC	Opposite sides of a parallelogram
2.	AC = AC	Common
3.	AB̂C = CD̂A	Opposite angles of a parallelogram
∴	△ABC ≡ △CDA	S∠S
∴	AB = DC	Congruence
	and AD = BC	

What is wrong with the reasoning? Of course: The pupil makes use of the theorem which he has to demonstrate, while he is demonstrating it. Even if the remainder of your demonstration is correct, the demonstration is nevertheless incorrect because in your demonstration you presume what you should prove.

The following type of mistake is one my pupils make ad nauseam for one reason or another – until the "errors file" has served its purpose (read more about this on page 78).

Incorrect observation and expectation

Look at the next two examples:
1. $2(m + 3n)^2 - 4(m + 3n) + 4$
   $= m + 3n - 4 + 4$

   The reason for this mistake is probably that you anticipate so strongly that this sum should factorize in the above manner, that you develop a blind spot for your own mistake. You want to use the "common factor" – and then you do it rashly.

Here is another example:

2.
$$\frac{x + y}{x + z} \cdot \frac{y - z}{y - x} \cdot \left(1 - \frac{z^2}{x^2}\right)$$
$$= \frac{-1}{x + z} \cdot \frac{y - z}{1} \cdot \left(\frac{x^2 - z^2}{x^2}\right)$$

In this case the pupil simply left out certain aspects of the sum which did not fit in with his expectations. When I pointed it out to him, he was surprised at his mistake, even more so than I!

The following mistakes (sometimes made by primary school pupils) also belong here:

$$\begin{array}{r} 39 \\ +24 \\ \hline 126 \end{array}$$

This pupil tells me "his teacher adds up like this!"

Probably he interpreted (understood, observed) her addition wrongly. The point I am trying to make, is that by *talking, explaining to me what he did, he could help me to help him. Talk about your mistakes, be prepared to say what you are doing – no one will ever laugh at you or think that you are stupid.*

That brings me to a particular evil: copying someone else's work. This is actually more than a mistake – it really is unforgivable, apart from the fact that it is dishonest and unethical.

Copying another pupil's work

Let me make it clear immediately: A friend who allows you to copy his work is doing you a great disservice. Whatever you do, *never, but never, copy someone else's work randomly – regardless of your motive, whether you cannot do the work and are afraid of your teacher's reaction if you do nothing or whether you "do not have the time to do the sum yourself even though you understand it", it does subconscious harm.* Besides, consciously you get the impression that you do not have to bother with your mathematics and that there is an easy way out. The result will become evident in the exam room, when you are sweating over the problems you never learnt to solve yourself.

Copying is the bedfellow of marking incorrect work as correct.

Marking incorrect work as correct

1. $\dfrac{a^2 - b^2}{bc + ac}$

   $= \dfrac{\cancel{a}^2 - \cancel{b}^2}{\cancel{b}c + \cancel{a}c}$ He actually cancels out the a and the b in the numerator with those in the denominator!

   $= \dfrac{a - b}{2c}$ And yet he marks his sum right!

Have a look at the following examples from Peter's book. He is a standard 7 pupil:

He who errs in mathematics and does not set it right, will err again!

And the next two:

2. $6x^2 + 19x + 10 = (6x + 2)(x + 5)$ (He marked this wrong sum right too!)

3. $2a^9 - 2ab^8 = 2a(a - b)(a + b)(a^2 + b^2)(a^4 + b^4)$

Of course the answer to 3 is correct – but the fact that there is no indication of the steps in between indicates only one thing (especially with this pupil): He has no idea how to do the sum and has copied it from someone else.

A pupil who does this sort of thing, is certain to have problems soon.

The second last problem which I want to deal with in this chapter, I shall call *interference*.

Interference and linearization

Interference means that a calculation or rule which was correct and relevant in a previous case, interferes with the current sum where it is not (fully) relevant (Steenkamp).

1. Simplify: $\dfrac{x^2 + 4x + 3}{x^2 + x - 6}$

$= \dfrac{1 + 4 + 1}{1 + 1 - 2}$ or $\dfrac{4 + 1}{1 - 2}$ In consequence of,

$= \dfrac{6}{0}$ $= \dfrac{5}{-1}$ for example, $\dfrac{4x^2}{2x} = 2x$

$= -5$

2. Simplify: $\dfrac{x + y}{\frac{x}{y} - \frac{m}{n}}$

$= \left(\dfrac{x}{1} + \dfrac{y}{1} \right) \cdot \left(\dfrac{y}{x} - \dfrac{n}{m} \right)$ In consequence of,

for example, $\dfrac{x}{\frac{x}{y}} = \dfrac{x}{1} \times \dfrac{y}{x}$

3. 0,581 is larger than 0,91   In consequence of the fact that 581 is larger than 91.

4. $824 \div 8 = 13$ Because you have been taught that 0 means "nothing", you want to ignore it when you do division.

There are even worse horrors, such as the following:

5. $\sqrt{a + b} = \sqrt{a} + \sqrt{b}$

6. $\log(3 + x) = \log 3 + \log x$

7. $\cos(z - 45°) = \cos z - \cos 45°$

8. $(a + b)^2 = a^2 + b^2$

A previous rule or principle interferes with your calculation. In this case, the previous rule is that of the distributive characteristics of real numbers, maths which you attempt to generalize without insight (linearizaton).

9. $2^3 = 6$

Knowledge of $2 \times 3 = 6$ interferes.

10. $6y \div y = 6y - y$

Knowledge of ordinary subtraction interferes.

11. Determine the value of x from:

$\sqrt{x^2 - 3} = 2$

$\therefore x^2 - 3 = \sqrt{2}$

This is an example of regression – you want to transpose the $\sqrt{}$-sign (which is of course totally incorrect and senseless) without knowing how; and of interference – you want to transpose the abstract square root sign, in consequence of the rule: if $ax = b$, then

$$x = \frac{b}{a}$$

12. Sometimes the process even works the other way around, when new knowledge interferes with old knowledge: You know that $m + m = 2m$. Then you find out that $m \times m = m^2$ and suddenly you have $m + m = m^2$!

We can therefore rightly call this kind of mistake interference. What happens here? You know a principle well, learn new work, do not understand it well enough and then try to apply your customary action to the new problem *without understanding what you are doing*. We shall discuss this again later on in this book.

The last type of mistake I am going to discuss will be called, with reference to the work of Olivier, distorted meaning and a lack of estimation.

Distorted meaning and a lack of estimation

Look at the following two examples:

(a) If one litre of petrol costs R1,12, how much does 3 litre cost?
(b) If one litre costs R1,12, how much does 0,53 litre cost?

I must tell you that to my utter amazement, Bell found that 73% of the 13-year-olds who did (b), got it wrong, while almost everyone had (a) right. Olivier gives the following reason for this: From their experience with whole numbers, the children know that you have to multiply to get a bigger outcome, whence the fact that they could do (a). From their experience they also know that if you want to obtain a smaller outcome, you have to divide – whence the fact that in the second case almost everyone divided. It is clear that they never really mastered the meaning of multiplication and division.

Of equal importance is the fact that they had only to make an intelligent guess to obtain an approximate answer. If one litre of petrol costs R1,12 and 0,53 is just over half a litre, then you *know* in advance – without *any* trouble – that the answer should be more or less 56c. That alone should really have made them think when they obtained the absurd answers they did obtain! Believe me, it is very valuable to *estimate* your answer beforehand, whenever possible – just as important as testing your answer afterwards. Talking about testing: *Always* remember to test your answers – a quick test is an effective way of making sure that your answer is correct and that you have not obtained an absurd answer! What is more, your test tells you soon enough when an answer can be ignored.

Well, now you know more or less what kind of mistakes (or combinations of mistakes) most other pupils also make from time to time. At this stage I want to urge you once again: *Make sure that you know what kind of mistakes regularly appear in your work.* As soon as you have sorted them out, you can begin a "mistakes file" for these typical mistakes – and then you can actively begin to work on these mark gobblers. *Remember: The timely, proper search for and correction of mistakes is the twin brother of success in mathematics.*

Conclusion

In this chapter you have read about the various mistakes pupils make, as well as some of the main reasons for these mistakes.

If you want to become an ace at mathematics, you have to start working at it *today*. Please note! I *do not* expect you never to make mistakes; only that you should find a solution to each mistake that you make. In the next chapter we shall briefly discuss why people so often say that mathematics is important. We shall also look at other ways of motivating yourself to take this subject at school, and possibly even later on.

CHAPTER FOUR

Motivation in mathematics

Introduction

You have probably often heard the following statement: "A boy (or a person) cannot do anything without mathematics." Of course this weighty statement sounds serious enough to persuade any child to take the subject. However, there are more than enough reasons for taking this subject if you are at all able to – and on pages 48 to 50 you can read how we determine whether a pupil can and should take mathematics.

How important is the subject?

The importance of mathematics

You know the stereotyped remark: "He can do mathematics, he must be very clever." There are other kinds of intelligence too – of course a person can be very clever without having to take mathematics after standard 7. If, however, you consult the yearbook of *any* university, technikon or technical college, you will realize that *mathematics is by far the most important subject for admission to these institutions of tertiary education.* Whether you want to study medicine, engineering or even psychology at university; or do a diploma in engineering at a technikon; or whether you want to qualify as an artisan at a technical college – you need mathematics. Of course there are fields of study where you do not need mathematics – but you can be sure that they form a very small minority.

By the way: The shortage of mathematicians and natural scientists in South Africa is often discussed at a very high level – and it is being stressed more and more that *this manpower shortage can be relieved to a great extent if girls' prejudice towards mathematics can be overcome.* I hope the girls have been reading carefully!

Mathematics as an optional subject

The mere fact that mathematics is an optional subject at school, creates the impression that mathematical literacy is not required of all pupils. Teachers and vocational guidance counsellors bear an enormous responsibility in this regard, viz. to make all boys and girls more aware of career possibilities (also for women) as well as of the importance of mathematics for so many careers.

I shall try to supply answers to the question whether you should take mathematics. But at this stage, let me say this: I have so often seen a pupil who discovers to his bitter chagrin and self-reproach, after school or in standard 8, 9 or 10, that mathematics is absolutely essential for the career he has in mind. On the one hand, someone like Gerald said to me: "I dropped mathematics in standard 7 because I *hated* it. Besides, my results were not at all good enough – but now I need it for further study." (On enquiry it transpired that 67% had been "not at all good enough" for him!) On the other hand there was Andrew who "worked for a distinction" and took the subject in the standard grade, only to discover in matric that it was essential for him to pass mathematics at the higher grade to be considered for admission to medical school. All things considered, you, your parents and your teachers bear a great responsibility to see that you take mathematics up to the highest possible level – otherwise you may soon reap the bitter fruits of short-sighted decisions.[1]

> *Remember!*
>
> *Mathematics is the only subject which can be studied after matric in all three main fields of study: Human Sciences (BA), Natural Sciences (BSc) as well as Economics (BCom).*
>
> *It is mainly the lack of mathematics which can pose a threat to your career plans.*

1. Of course you should take mathematics in the higher grade rather than the standard, because:
   (a) It ALWAYS gives you a better base for further study.
   (b) It almost always improves your chances for admission substantially; for example: at a certain technikon an A in standard grade counts 59 points while a D in higher grade counts 61!
   (c) Mathematics in the higher grade is a REQUIREMENT for numerous fields of study. Therefore, do not switch from higher grade to standard grade for short-sighted considerations.

By this time you are probably thoroughly aware of the necessity of the subject and the need for persons educated in a mathematical field of study. But how do you motivate yourself to study mathematics?

Motivation

To start with: Always remember that at university level we do not refer to students as being clever or stupid at mathematics. We talk only of students who work hard at mathematics – and who do well; or of students who do not work hard at mathematics – and do not do well. I stand or fall by this: your will to do well in this subject, your perseverance, your self-discipline, your diligence and your hard work will eventually determine whether or not you will succeed in this very important subject – to a far greater extent than an arbitrary figure which is supposed to indicate your aptitude. The following lines are particularly applicable to achievers in mathematics:

> *The great heights reached by men*
> *and kept*
> *were not attained by sudden flight*
> *but they*
> *while their companions slept*
> *were toiling upward through the night.*

You may wish to bear this in mind.

> *The inside lane in the mathematics race depends more on your will and readiness to work hard every day than on mere talent.*

What do the experts have to say about motivation for mathematics?

The behaviourist[2] view of motivation for mathematics

This school stresses so-called external motivation. This means that your parents', your teachers' and other people's reactions to your efforts in mathematics will determine your eventual success

---

2. The behaviourists belong to a school of psychology which explains human behaviour in terms of reactions to stimuli.

or failure. If these people reward your achievements and encourage your efforts sufficiently, you will continue to perform well. This can take the form of praise, a star, positive comment or high marks. And if other people do not encourage or reward you sufficiently, reward yourself every time you have achieved something in mathematics: by doing something you enjoy; by being *proud* of your achievements; by always setting yourself new goals. Pat yourself on the back and *enjoy* your achievements – however small.

Piaget's view of motivation in mathematics

This pioneer in the field of mathematics and the psychology behind mathematics differs from the behaviourists. He sees motivation for mathematics as something which should preferably come from within oneself. He maintains that you will want to learn to do mathematics *if it makes sense to you, if you see it as worth your while to learn to do it.*

But I believe in a global approach to motivation.

A global approach

This means that there should continually be a balance between internal and external motivation. Placing undue and one-sided emphasis on one of the two can easily lead to secondary problems.

In other words: It is of the greatest importance that others should give appropriate recognition to your prowess in mathematics, however insignificant it may seem to you. But it is equally important that you should, as soon as possible, learn to work for yourself, to compete against yourself, to be clear in your own mind that mathematics is important to *you* and that you will do your very best in the subject for this reason.

One day I asked a large group of pupils to write down when they were motivated to do well in mathematics (not necessarily in order of preference). I received the following responses:

- when they enjoyed it;
- when they understood it;
- when they saw it as useful and meaningful;
- when they wanted to pass an examination;
- when they liked the teacher and his teaching methods;

- when they managed to do the sums and enjoyed doing mathematics; and
- when they received *immediate* feedback: in other words, when their achievements were rewarded immediately; when they knew immediately whether an answer was right or wrong; and they did not have to wait for weeks to receive the results of tests, etc.!

With which of these motivations do you agree? Can you tell which of them represent internal and which external motivation?

When I asked the same group when and why their motivation decreased, their answers were just as revealing:

- when there was little or no support from their parents;
- when they did not understand the language and symbols in mathematics;
- when the work was too abstract;
- when they did not understand the contents;
- when they were bored;
- when the mathematics seemed meaningless or irrelevant;
- when they always had their answers wrong;
- when the work was too difficult and they could not cope;
- when they did not like the teacher and his methods;
- when they had reading problems; and
- when they had been absent and had fallen behind their classmates.

Once again, see whether you can tell which of these responses represent internal and which external motivation. To what extent do you agree? Is this the case with you too? Is there something you would like to add?

A quick remark about the idea that some people hate mathematics: one should quickly decide that mathematics is a subject you can enjoy and like, instead of something you hate and have to endure.

What about competition?

Competition

To start with, always distinguish between the desire to be the best at mathematics and, much more aptly, the desire to *do your best*. Always remember that for every winner out of fifty pupils, there will be forty-nine losers! The desire to win at all costs often leads to the situation where only those who stand a realistic chance of winning want to take part; after a while the others do not really want to take part any more. Of course this does not mean that all competition is harmful or detrimental. Friendly competition can lead to improved performance and even to better friendship if you and your "opponent" can encourage each other.

Do you still remember the good old days when you were about four years old? How you used to play alongside other children, but rarely *with* them? At that stage you had not learnt to share and to co-operate. Your group activities were limited to having a rest together or to listening to a story together with other children. This applies up to the age of six. But as you grow older, and especially when you reach around standard 6, you do want to be part of a group. For this reason I recommend that, wherever possible (and especially when you have problems), you should do mathematics *with others.* Discuss your problems, difficult concepts, anything you like, with persons of your own age and peer group. Such a discussion can be an inestimable motivating force if used correctly.

Psychological studies have produced these interesting findings: Boys (especially from standard 5/6) want to move in *groups* while girls prefer *pairs.* The idea of being right or wrong is therefore a more important and sensitive matter to girls than it is to boys. This is another reason why girls prefer to drop mathematics (where they can often get the idea that they are wrong) when they have the option in standard 6. Now that the girls are aware of this, I want to plead with them *never to make a short-sighted choice for the wrong reasons.* If you are in doubt as to whether you may/should/can/want to/must/take this subject, *please* go and talk to an expert in the field.

Does being interested in mathematics play a role in your motivation for the subject?

Interest and motivation

Behaviourist researchers often deprived animals of food or hurt them in order to obtain a reaction. But unlike animals, children are often not interested in sensations of hunger, thirst or pain. When a child is absorbed in something (for example solving an interesting problem or in an interesting game) he will even ignore external diversions such as a cut to his finger or leg, in order to finish his game or assignment. In other words: if you are sufficiently interested in something, you will be highly motivated to do it to the best of your ability. As regards your interest – or lack of it – in mathematics, the following matters are very important:

- There is a very slight connection between your aptitude for mathematics and your interest in the subject. In other words, it is simply untrue to say that you have no interest in the subject, only because you supposedly have no aptitude for it.
- Emotion does play an important role in your interest in mathematics. This simply means that your interest will obviously increase once you begin to achieve success in it.
- Knowing your subject and knowing that you can do well in it will largely determine your interest in it. By this I mean that it is short-sighted to say that you know in advance that you will not like the subject (or be successful in it) unless you have first made a conscious effort to remove obstacles in your way.

You have probably heard the expression: "Take the subject for the sake of its effect on you, rather than your use for it." What then are the true goals in taking the subject?

Goals in taking mathematics

There are mainly two standpoints in this regard and they differ quite widely.

The first standpoint is that a pupil should take mathematics for the sake of its formal formative value. This means that the subject develops your mind to such an extent that you will generally achieve more success in your other subjects and in other fields in life.

The second standpoint is that the subject should mainly be taken for the sake of its materially formative value. This means that you should take mathematics for the sake of the knowledge and skills which it teaches you. This implies that you take the subject in the first place for the sake of the doors it opens for you, doors which would otherwise stay closed.

As stated before, I believe in a global approach – both standpoints contain elements of the truth. A few of the general objectives in taking mathematics might include the following:

41

- Mathematics is a powerful instrument which allows you to communicate with others; the use of the language of mathematical symbols is indispensable to any society (for example, imagine a world without figures!).
- Mathematics is a vital, supporting subject (in fact it cannot be ignored) for other sciences. (Why do you think that the majority of fields of study require matric mathematics?)
- Mathematics has an infinite number of indispensable applications in industry.
- Mathematics enables man to understand, order and influence his environment.
- School mathematics enables you to pursue the career of your choice and possibly to continue with it at a later stage or with a related science.

Do you realize to what an extent *all* subjects touch upon mathematics? Try and tell which of the following mathematical operations you do in your other subjects:

1. Counting: 1, 2, 3 ... 100, ... 300, ...
2. Arithmetic: 34 + 5; 45 × 5; 67 − 8; ...
3. Drawing patterns (think of the patterns for clothes, for example in home industry).
4. Arranging things in groups (like animal species in biology).
5. Drawing forms (like rectangles, circles, triangles).
6. Making models (for example a cube).
7. Drawing and interpreting graphs (as graphic representation of the gold price in economics).
8. Dealing with money matters (as in economics).
9. Working with decimal fractions (as in domestic science: 0,5 kg flour).
10. Working with surface, volume, length (as in natural science).
11. Using fractions (as in physics).
12. Measuring things (as in woodwork).
13. Weighing things (as in home industry).
14. Estimating or guessing numbers (as in geology).
15. Use of time (as in history).
16. Drawing objects or maps to scale (for example 1:2 000) (as in the Army, when you go on manœuvres).
17. Drafting tables (as in Afrikaans and English).
18. Measuring temperature (as in geography).

The list is unending. Never again sit in your history class thinking that all this has nothing to do with mathematics!

Conclusion

In this chapter you have learnt more about the link between achievement in mathematics and motivation. I have tried to give you good reasons why you should really do your best in the subject by showing you what mathematics can do for you – and what it can mean to you as a person.

In the next chapter I shall discuss the question: How do I know that I can or should take mathematics?

CHAPTER FIVE

Can I take mathematics? Should I take it?

Introduction

I want to start this chapter with a quotation from Gannon and Ginsburg: "Nearly all children dispose of the informal abilities to achieve success in school mathematics. Most problems with school mathematics flow from social, emotional, personality and educational factors." Or as Kumon puts it: "Practically every child can be a wizard in maths." Although this is a debatable point, I am thoroughly convinced that if we take the necessary steps, we can drastically improve the success rate in mathematics, at school and afterwards. *And that the largest common denominator* among the poor performers in mathematics *is plain laziness!*

First, let us briefly look at the latest research findings about the functioning of your brain.

Your brain and brain dominance

You may have heard that Einstein said that the normal person barely uses 15% of his total brain capacity during his lifespan. This figure is arbitrary – but you can assume that we use a very small portion of our true capacity.

More recent research has specifically explored the role of the left and right hemispheres of the brain. It would appear that the left hemisphere controls your verbal (language), numerical or figure and logical functions. This hemisphere specializes, as it were, in abstract or symbolic representation, where the symbols do not have to have any physical resemblance to the objects they represent. The right hemisphere apparently controls your spatial (high/low), visual (how you see), perceptual (observation), intuitive and imaginative functions (including your creative abilities, and emotions like sorrow). The right hemisphere contains repre-

44

sentations which resemble (i.e. which are isomorphous to) reality. Therefore, when you think, write, read and listen, you use the left hemisphere of your brain. This is probably very good for a child with a dominant left hemisphere *but* it creates problems for a child whose right hemisphere is dominant. Children with a dominant right hemisphere must be able to see, feel, touch, imagine, manipulate, taste(!) and use their "sixth sense" (which is probably only a question of "programmed reaction") in mathematics! The behaviourists say: "There is nothing in the mind that was not first in the senses!" All of this means that the two methods of learning or teaching (abstract explanation or concrete learning) are equivalent and complementary.

Once again, I propose the global approach: that, as far as possible, you perform all the actions that I have mentioned above when you do mathematics – whether it is in the classroom, at home or wherever. If your teacher does not allow all these actions, do ask him to accommodate you – it may mean the difference between success and failure. It is a well-known fact that Western teachers neglect the right hemisphere shamefully – much can be done to create opportunities for a pupil to visualize and to use his imagination.

As the principle of visualizing (or visual representation) may be of great significance to certain pupils, I quote two examples to illustrate what it means.

Example 1 (from an article by Norma D. Presmeg)
A dog is chasing a fox which is 30 metres ahead of him. The dog's leaps are two metres long while those of the fox are one metre long. For every two leaps the dog makes, the fox makes three. What distance will have been covered by the time the dog catches the fox?

Look at the following sketch by a boy in standard 8:

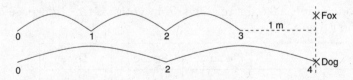

Based on this he simply reasoned as follows: You see, for every four metres covered by the dog he catches up with the fox by one metre. In other words, he catches up

$$1 \text{ m in } 4 \times 1 \text{ m}$$
$$\therefore 30 \text{ m in } 4 \times 30 \text{ m} = 120 \text{ m}.$$

Of course the problem may be solved without a sketch and the method you use is a matter of personal choice. There is no doubt, however, that a sketch or a diagram often helps.

Example 2: (from the final matric examination paper of 1990 – a sum which very few pupils managed to solve owing to its supposed degree of difficulty!)

It takes an aeroplane $\frac{5}{6}$ of an hour longer to cover a distance of 1 000 km when flying directly into the wind, than when flying directly with the wind. If the wind maintains a constant speed of 100 km/h, determine the normal average speed of the aeroplane in the absence of wind. Look at the following scheme (it is crystal clear that you are going to work with distance, speed and time – therefore keep your formula sketch handy):

$$D = S \times T \quad \ldots\ldots (1)$$
$$S = \frac{D}{T} \quad \ldots\ldots (2)$$
$$T = \frac{D}{S} \quad \ldots\ldots (3)$$

To begin with: What is being asked? Speed, which we shall take as x km/h.

Firstly, draw a graphic representation of the two flights:

Aeroplane speed: $x + 100$ .... (4)

Wind direction $\Rightarrow$ $D = 1000$ km

Aeroplane speed: $x - 100$ ... (5)

Which elements are missing from our $\triangle$? Time, of course!

Now draw up a table in which you interpret all the information. Simply take the readings from your sketch. Transpose (4) and (5) into (3):

	Distance	Speed	Time	
With wind	1000	$x + 100$	$\dfrac{1000}{x + 100}$	-----(6)
Against wind	1000	$x - 100$	$\dfrac{1000}{x - 100}$	-----(7)

Finally: State the equation which has been given "implicitly", in your own symbols:

$$T_{\text{with wind}} + \frac{5}{6} = T_{\text{against wind}} \cdots\cdots (8)$$

> *Try to link your work logically, whenever possible, by means of references like (1), (2), (3), (4) up to (8) above.*

There you are, now you can replace (6) and (7) in (8) and just solve the equation:

$$\frac{1\,000}{x + 100} + \frac{5}{6} = \frac{1\,000}{x - 100}$$

$$\times\ 6(x + 100)(x - 100)$$

$$\Rightarrow 6(1\,000)(x - 100) + 5(x + 100)(x - 100) = 6(1\,000)(x + 100)$$

$$\Rightarrow 6\,000x - 600\,000 + 5x^2 - 50\,000 = 6\,000x + 600\,000$$

$$\Rightarrow 5x^2 - 1\,250\,000 = 0$$

$$\Rightarrow 5(x^2 - 250\,000) = 0$$

Who would say that mathematics isn't fun! I honestly think I can manage to explain this "very difficult" sum to a standard 1 pupil.

$$\Rightarrow (x - 500)(x + 500) = 0 \quad \text{Ignore} \quad (x + 500) = 0 \quad (\text{why?})$$

$$\Rightarrow x = 500 \text{ km/h}$$

With repeated practice in visualizing, schematization and tabulation become really easy and help you to enjoy your mathematics.

To conclude: When doing mathematics I suggest that you read (*aloud* if you like), write, think, use your imagination, *use as many of your senses as possible*, make as many sketches as possible (that includes graphs of any kind), draw schemes, look at models – *and that you do it actively and with dedication*. But we shall discuss this again later on.

Is it only the so-called "intelligent" child who can do mathematics? I am certainly not going to inundate you with answers to this question – the answer is a short and sweet NO.

Mathematics and intelligence

It is generally accepted that "it is only the intelligent child who can make a success of mathematics". This statement is highly controversial. In the first place, the definition of intelligence, as we know it, is utterly debatable. I know every child is extremely curious about his IQ – but let me impress upon you that *no person or child "has an intelligence coefficient"*.[1] Your IQ is merely derived from a series of mental tasks put to you at a certain time by a fallible person, not something which is fixed at your time of birth and can never change (like the colour of your hair and eyes). At another time and with someone else you would probably have a different score! It does give a reasonable indication of your general intelligence. I dare say that any average pupil who reaches standard 6 has the ability to do mathematics – regardless of his IQ.

Research in this regard is often extremely confusing and frustrating. One researcher might say that the so-called verbal IQ has little connection with mathematics (despite the fact that the verbal part of the individual IQ test actually includes the mathematics test), while another maintains that the non-verbal part has very little to do with mathematics.

I honestly think that you can safely assume that if you obtain reasonable marks for mathematics in standard 5, you are intelligent enough to carry on with the subject. And if this is not the case, you still have to remember that factors such as your family circumstances, your health, your interest and many other factors probably played a part – factors which you can control.

At this stage I want to issue a warning to those pupils who undergo a test somewhere and then return with the smug remark: "They tell me I can follow any career." Remember that you cannot do anything at all if you do not work very, very hard – and especially at your mathematics, if you are at all interested in a field where mathematics is a prerequisite. In my profession I see far too many pupils and students who pride themselves on their "gift" for the subject, but who fail it or do so badly in it that they cannot gain entrance anywhere. And the promise that "... I shall repeat mathematics to obtain higher marks in matric", *only works in highly exceptional circumstances. The right (and I would like to say the only) time to get your marks in order, is now – not later.*

1. In the book *Make your child brighter* I discuss this matter exhaustively.

At this stage you might read on page 13 what I told your parents in this regard.

To summarize this part: If you can score 50% in mathematics in standard 5 (and I think most pupils are capable of this), then you can also pass matric mathematics. However, just as you cannot swim in a swimming pool without water, in the same way you cannot become an achiever at mathematics without the necessary input.

Creativity and mathematics

To start with I *always* look for the "wild ducks"[2] in my mathematics classes – the original pupils who ask questions, are not afraid to ask *anything* (even though their friends might think that they are stupid), who let their imagination loose, who are prepared to take the plunge, who try other solutions, who are ingenious, who think *differently*. They are often the most successful, because they believe that there is a solution to any problem, even the most difficult one in mathematics. Like John who told me one day, "I know the chaps think I curry favour with my teacher, but I don't care, I always try everything to think and do a little differently from the others and it works, I am the *star* in our class!" Krutetskii differentiates between *scholastic ability* and the ability to do mathematics creatively. The former refers to your ability to master mathematical facts and to render them, whereas the latter means that you have to be able to create something original. He certainly has a point. *To my mind one can assume that the ability to become, for example, a successful student in engineering, depends to a great extent on your ability to be creative in mathematics.* A good mark in geometry is an indication of this ability. But I want to reassure you, if you do not have a great measure of this, you can still gain a distinction in matric, even at higher grade, if you work hard enough.

Just before the finals of the tennis championships at Wimbledon a journalist asked Boris Becker what his strategy for the finals would be. "That's easy," was his reply, "I shall try to

2. Those pupils who are prepared to try something different, to approach a problem differently, who ask questions and do not just take things for granted – like John who wants to know whether "multiplication is not merely repeated addition", or Peter who approached 252 + 117 as follows: 200 + 100 = 300; 50 + 10 = 60; 2 + 7 = 9; answer: 369!

win the last point ..." That is practical, realistic creativity at its best – the kind I am looking for in the mathematics class. Let me give you a few important examples of mathematical creativity.

1. Always look at a sum or matter from different angles, from different viewpoints. Look at the picture below. Do you see what I mean? If you only look at the black parts, you see two faces facing each other. If, however, you look at the white part, you see an hour-glass. Also look at the questions on pages 81 and 82. *Always observe a problem from different points of view, especially when it seems to you that you have reached a dead end.*

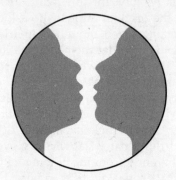

2. We all know that 3 is exactly 7 less than 10. But there are different ways of getting from 3 to 10:
   - $3 + 1 + 1 + 1 + 1 + 1 + 1 + 1 = 10$
   - $3 + 7 = 10$
   - $3 \times 3\frac{1}{3} = 10$
   - $3 + 5 + 2 = 10$
   - $3 + 16 - 9 = 10$

   Of course, there are many other ways in which one can obtain the result. It is usually wise to take the shortest path to the answer (*draw a red triangle around the shortest path here*). But very often you have to follow a "different" strategy to solve the sum; you have to be a little creative.

3. In the third and last case, I want to illustrate another elementary example of practical creativity, the kind of creativity of which you have a special need in mathematics.

   Read the following problem and try to reach the solution *yourself* – use any sketches you might consider necessary!

You have 13 bags full of wheat. All the bags weigh 50 grams, except one which contains chaff with the wheat. This bag weighs 45 grams. There is no outward difference between the bags and you have a 50 gram counterweight which you may use only *once*. How will you determine which bag contains the chaff?

(Solution: Arrange the bags in pairs and compare their weight by placing them on balancing scales. If you find a pair where one bag is lighter than the other, you know that that bag contains the chaff. If all six pairs balance, you know that the remaining bag contains the chaff!)

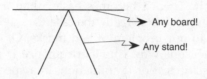

Of course there are other ways of solving the problem – name a few. The main point I want to stress once again is that *practical, elementary creativity* is of great value in mathematics, especially in geometry.

This chapter would be incomplete without the vital question: whether bad marks in one or two tests or examinations mean that I have suddenly become a dunce.

When does a bad mark indicate a serious problem?

I often hear the following arguments:
- I am now lagging behind in mathematics: standard 9 mathematics is totally different from that of standard 7 or 8 – I am not clever enough to do maths this year.
- I have now had poor results in two consecutive tests (or in a specific examination); therefore I probably have a serious problem with the subject.
- I thought I was writing so well; when the results came out, I found that I had really done badly.
- I understand the work, but I still do not get good results.
- I never seem to finish mathematics – and I always lose a lot of marks as a result.
- I can still manage algebra, but geometry is too much for me.

The common theme is: There is a chink in my mathematics armour. I have been expecting a serious problem all along, now at last I have the proof I have been looking for. I am not really clever enough to do mathematics.

Let me reassure you. Nearly all of us have this tendency – from the very best students, to the chaps who struggle. I hear the same story repeatedly. It forms part of human nature. We are fallible and when we allow doubt to sneak in, it gains a foothold; especially where a difficult subject like mathematics is concerned. Now that I have used the word difficult on purpose, just the following: a really famous professor said to me many years ago that all mathematics is extremely difficult until you understand it. Then it becomes easier than any other subject. Take it or leave it, that is really the case.

But back to our problem of bad marks. Of course, there can be various reasons why your performance is poorer than normal in a specific test or examination, in a specific section, on a specific day, under a specific teacher. Let us discuss a few possibilities – you might just recognize yourself here.

1. Henry was referred to me because he had failed his mathematics hopelessly. However, tests indicated that he was certainly not the dullest pupil I had ever come across. I did notice that when I allowed it, he spent far too much time on individual tasks/sums. He would check every step again and again. Of course he could never even come close to finishing his assignments, tests and examinations. Besides, his perfectionism (he could not distinguish the essential from the non-essential and was the type of pupil who would fill pages when it was possible to complete something in a few lines) made him very anxious, not to mention the accompanying frustration. I had to teach him to spend his time more effectively and to use more efficient methods of solving problems. After that, he gradually regained his self-confidence in mathematics.

2. Gerald's mother was desperate. He just could not get on with his teacher and, on top of that, he failed mathematics dismally. After close investigation, the following transpired:
   • Gerald's books had not been marked by his teacher for weeks; incorrect sums had been marked correct, corrections had rarely been done. When I corrected one of Gerald's mathematics tests, I awarded him 27 marks out

of 40 – while his teacher had given him 12 out of 40. How can this be possible? I repeat: we are human and humans err. This difference of 37,5% is really surprising but, believe me, it happens every day. With time, and once the problem with his teacher had been settled in a mature fashion and Gerald had improved his input drastically, his mark in mathematics returned to 75%.

3. Harry performed poorly in mathematics and he was extremely reluctant to even attend extra classes in mathematics. "We are just wasting my father's money and time even further," he would say regularly. In effect, I really did not achieve anything with the lad, until one day he, his parents and I had an open-hearted discussion. (I had suspended his classes.) It then came to light that this boy was so worried by his parents' frequent quarrels that he simply could not concentrate.

4. Betty was a particularly pleasant girl who suddenly, on two consecutive occasions, obtained 40% less than her normal marks. When her mother brought her to me and after we had simply had a good chat, she said with a gleam in her eye, "I will show you that I can" – and she kept her word. At the next occasion she had regained her old mark of 70%. The reason? Good old teenage love which disrupted her work.

5. John was one of those prodigies as far as the PC and the pocket calculator were concerned. He could do the cutest things with his PC and could do the most complicated calculations with his pocket calculator. But it was useless to ask him to execute long division somewhere – he was practically addicted to his pocket calculator. To cut a long story short: despite his prowess with the computer, he was fast losing ground in mathematics. I am of the opinion that a pocket calculator and a personal computer can be *aids (valuable aids)* – but that they should *never* be allowed to take the place of knowledge of the basic principles. As Finn puts it in the *Wall Street Journal:* "Like with so many previous efforts at reform in the past, American educationists, together with the 'experts', show all the classic signs of walking over the edge of the cliff like lemmings,[3] taking their

3. Lemmings are small Arctic rodents, one species of which migrates when their population reaches a peak. Large numbers have been known to hurl themselves into the sea during these migrations in what appears to be a mass-suicide.

pupils, pocket calculator in hand, with them only to tumble into the sea of mathematical helplessness."

It took me many hours to bring John's basic mathematical insights up to date and to cure him of his over-dependence on his pocket calculator.

What am I trying to get across to you? That one must *always* look for the problem behind the problem. As Piaget said: We must *always* find the *underlying* problem; the thought patterns or other factors that cause problems. You can read more about this on pages 100 to 104. But at this stage you can start by remembering the following *golden rule:*

Failure in mathematics is a symptom, not an ailment. Whether it seems serious or less serious, we must always look for the reasons behind the fact of failure, and we must approach each case individually.

Conclusion

In this chapter you have read why I maintain that most pupils are capable of taking mathematics up to matric level. I am not saying that you should become a professor in mathematics. What I am saying is that you are intelligent enough to take as much mathematics as you may need to pursue your specific line of study. I have also pointed out the important fact that failure in mathematics is an octopus with many tentacles – and that you should always ask somebody to help you look for the reason behind your specific mistake or mistakes.

In the next chapter you can read more about the phenomenon of mathematics anxiety, how to concentrate more fully, and the significance of humour in mathematics.

CHAPTER SIX

Mathematics anxiety, concentration and humour

Introduction

Since Sheila Tobias wrote a popular article in a women's magazine about mathematics anxiety and followed it up with her book *Overcoming math anxiety*, in 1978, people have begun to talk more readily about their own fears with regard to the subject. There are few of us who have not experienced the helpless, unsettling, almost panic-stricken feeling after being confronted by an apparently insoluble mathematical problem.

Because it is such an important factor that it could influence your choice of subjects and even your choice of a career, in fact your marks in mathematics, I want to discuss it briefly.

Mathematics anxiety

Although it would be more correct to refer to a pupil's negative attitude towards mathematics or to refer to fear of mathematics, I shall use the term mathematics anxiety here, because it has already become established in popular usage.

Does the following article from *Sarie*, 18 March 1981, sound familiar?

Mathematics Fear

The standard 7 son of Mrs M. v. R. in N is doing increasingly poorly in mathematics although, for more than a year, he has been taking individual private lessons at great cost with a particularly able woman in a neighbouring town. He has now developed such an irrational fear of mathematics that he has the most terrible nightmares before writing mathematics examinations. Mrs M. has found him screaming and in a cold sweat in his bed at two o'clock in the morning. He would like to become a dentist and matric mathematics is a prerequisite for this.

A few pupils develop this fear early on while others seem to make a satisfactory beginning with mathematics and then develop such an aversion to the subject during their high school years that they reach the point in their mathematical education where they cannot or do not want to carry on. It is understandable that a pupil (like the boy mentioned above) who knows

that he needs the subject for his future studies but who cannot do well, should develop an intense aversion to mathematics.

According to Visser, the causes for this phenomenon are to be found mainly in primary and high school. Children all too often learn principles and rules without the slightest insight and at some point during their high school career, they reach the point where simple memorizing is not sufficient.

Add to this the fact that every pupil knows full well that, unlike most other subjects, he cannot start from the beginning every year and you have a recipe for anxiety. A pupil who misses out on a principle or two, or who performs badly during one year and who does not actively work very, very hard, is simply not going to make up his backlog. Incidentally, American research indicates that up to 70% of all mathematics students suffer from mathematics anxiety, while 11% of them suffer from it so seriously that they need urgent counselling.

I want to go as far as to say that the teacher who allows his class to ask the "dumbest" questions and who *never* makes his pupils feel stupid, holds a mighty weapon against mathematics anxiety. For this reason, it is of the greatest importance to me that you should have so much self-confidence and such a good self-image in mathematics that you have the pluck to ask questions about aspects of mathematics which bother you, at any time, in any place, of anybody. You might remember in this regard that mathematics teachers are not cruel or sadistic, but that they are honestly not always aware of the fact that mathematics can cause such terribly profound emotions in you. *Despite the fact that pupils sometimes experience their teachers as severe, aggressive or unfriendly, it appears from research that this subjective perception is actually unfounded!* Rest assured that you do not need, as Buxton puts it, to keep saying, "Thank you, thank you for not being annoyed", when you do not know an answer or when you give an incorrect answer. But please do see that you do your humble part to the best of your ability at all times.

You have probably heard the following words often enough:

Help, I "struck a blank"!

Before I go any further: a small measure of anxiety is always desirable when you are doing a test or an examination. This not only increases your preparedness, but also leads to higher achievement. But not so much anxiety or fear that you "strike a blank"!

I "struck a blank"!

Is it possible not to remember a thing, suddenly? According to most students in my classes, it is. And how do you handle this? Rather, how do you prevent anxiety from catching up with you at some point? Let us discuss a few guidelines.

1. Learn a few relaxation techniques. Get the help of someone like your guidance counsellor or sports coach for this. The following are a few such techniques:
   - Adopt a comfortable position. Lie on your back, or sit comfortably. If your clothes are close-fitting, loosen them. Breathe deeply and hold your breath for five seconds. Now breathe out and experience the feeling of calm which starts to take hold of you. Then breathe normally and concentrate on the pleasantly heavy feeling which takes hold of your body.

- Now contract every muscle in your body. Start with your face, jaws, shoulders, chest, back, stomach, arms, legs, hands and feet – contract every part of your body. Feel the tension throughout your whole body. Then allow *all the muscles to relax completely* and experience the feeling of calm coming over you.
- Note the difference in the tension you experience when your eyes are open and the relaxed feeling you experience when your eyes are closed. Now relax all your muscles once again, keep your eyes open and feel how the tension leaves you as soon as you close your eyes.
- Keep your eyes closed, breathe in deeply and hold your breath. Keep your body relaxed, but be aware of the fact that you are still tense because you are holding your breath. Then let your breath out and experience the deep, relaxed feeling that takes hold of you.
- Breathe normally, and experience the warm feeling of relaxation which flows through you. If you completely surrender to the feeling of relaxation, if you experience it in every part of your body, you should feel calm, heavy and secure. Remember: Complete relaxation leaves you with a feeling of peace and security.

    If you really cannot relax, you may want to consult your doctor; but always remember to test any medication timeously in order to avoid any possible side-effects.
2. See your doctor if you really cannot rid yourself of your anxiety.
3. Re-evaluate your examination or test situation. This means that, in the first place, you should change your attitude towards the situation from negative to positive. Always remember that study without evaluation would serve no purpose; that you actually get the chance to evaluate your knowledge of and insight into mathematics; that you can determine where your weaknesses and problems lie; that poor results do not mean *that you are a failure and will never be able to pass.* You are always precious to those close to you irrespective of your achievements. Results have no bearing on you as a person, you should never forget that.
4. Set yourself realistic goals. Really take the trouble to find out, if necessary from a psychologist, which goals and aspirations are realistic in your case and abide by them.
5. Develop good study methods. In the first place ask your coun-

sellor to help you or ask your parents to look for help else-where or to buy you a good book on the subject. Also read what I have to say in this regard on pages 81 to 87.

I often refer to the significance of concentration for achievement in mathematics.

Concentration and attention

At the outset: When you are doing a sum, you have to concentrate. When you are listening to a number of facts and you try to remember them, you are paying attention. In other words, in order to concentrate you have to be *active,* and to pay attention you have to be receptive and yet also *passive.*

These are key concepts in mathematics. A person who cannot concentrate will probably experience problems, serious problems, in mathematics. Paying attention is not enough. Mathematics essentially requires that you be *active* – that you write, think, do, propose, use your imagination, fantasize, make sketches – never just *sit and do nothing.* Problems simply are not "figured out" passively, you *must* as far as possible involve your total brain, your whole person in it. Bloom and Broder rightly say: "The big difference between successful and unsuccessful students in mathematics lies in the extent to which their approach to mathematics can be seen as *active or passive.*"!!

But everything in class should not be all seriousness. No, indeed, one should be able to introduce some fun into slogging!

Mathematics and a sense of humour

Research shows without any doubt that pupils perform better in and give preference to classes when their lecturer or teacher shows signs of a sense of humour. It creates a positive atmosphere in class and " just allows you to relax", according to a pupil. It gives you the confidence to ask questions, to take up challenges (so terribly important in the mathematics class). It helps you to adapt, it improves your self-image. In fact, it helps you to unbend, to see yourself in a different light. *The development of a sense of humour, the ability to laugh at yourself, undeniably adds to a better self-image; to more self-confidence.* Allport has a point when he says that a pupil who can laugh at himself, can solve his mathematics problems. Even your most painful experiences in the mathematics class can take a humorous turn. It will enable you to face up to your personal and emotional problems again. You do not have to go so far as to make fun of yourself – what I am saying is that by laughing when necessary, you can relieve tension, feel more at ease socially, see your mathematics problems in perspective again, avoid exaggerated seriousness, make yourself felt in a different, more effective way and at the same time counter mathematics anxiety effectively.

Conclusion

In this chapter you have read about mathematics anxiety, concentration and attention; and you have received a few guidelines for countering this anxiety.

In Chapter 7 you will read more about the relation between mathematics and language, self-image and self-confidence. We shall also discuss the important question of whether you should listen to music while you are doing mathematics.

CHAPTER SEVEN

Mathematics, language, self-confidence and music

Introduction

Goethe once said the following about the language of mathematics, "Mathematicians are just like the French. Whatever you say to them, is translated into their own language ... and then becomes something entirely different!" Mathematics indeed has a language of its own – a language you can and should master by sheer hard work, otherwise the subject will remain inaccessible to you.

That is not what I had in mind when I said you should borrow when doing division!

Mathematics and the language of mathematics

Mathematics has a limited, technical language which is part of your mother tongue.[1] Because it is a *language*, somebody who is good at languages or finds it easy to learn a foreign language also has the ability to learn mathematics and I stand or fall by this. Terms that do not otherwise occur in English are seldom used in mathematics. It is true that words like logarithm, sine, cosine and tangent are specific to the subject, but these are examples of the few really foreign words in mathematics.

However: although the language of mathematics is composed of normal English words, these words are used in a specific and unchanging fashion. Ordinary words may sometimes take on other meanings, but the meanings of words in a mathematical context remain unchanged.

This means, among other things, that the possibility of word play and misunderstanding always exists. Let me explain:

1. When I asked a group of young children (11 to 12 years old) in a mathematics class for the meaning of the word *volume*, most of them replied that it was the knob on the television set, which you use to make the sound louder or softer!
2. Bentley and Malvern quote the following gem which says it all: A boy in standard 6, 13 years old and not at all stupid, kept referring to "tree dimensional". When someone asked him what he meant, his reply was, "My teacher always draws the following tree when he talks about tree dimensional" and he drew the following sketch:

Language problems

Pupils who have problems with mathematics, very often have language problems. They often have problems with general terms in the vocabulary of mathematics, like the following:
- up/down/in(side)/out
- plus/square number/square root

---

1. According to Bell, a child has to read only 300 to 400 new words in primary school and 100 new words a year at high school in order to master mathematics as a LANGUAGE.

- the square root of the sum as opposed to the sum of the square roots
- real/exponential

If you feel that you have problems with any of these, it is high time that you start making a note *whenever* you come across one of them. Keep your notes handy so that you can look at them again and again – that is the only way you are going to master them. Your teacher should be able to help you make an exhaustive list of the essential terms if you ask him. Use these terms when speaking to people, explain them to someone as soon as you understand them yourself. We find that pupils who struggle with mathematics often find it difficult to communicate that which they do understand. Ask your teacher to keep his language as simple or uncomplicated as possible.

In addition to this, there is the problem of your position in space.

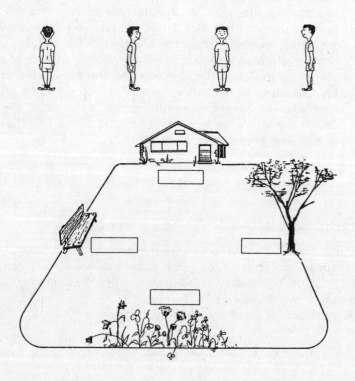

1. Place the boys in the middle of the sketch one by one.
2. Write "right", "left", "in front of", "behind" in each rectangle (in relation to each specific boy).
3. Now complete the following:
   The bench is ....   The flowers are ....
   To the left of the boy is ....   Behind the boy there is ....
   The house is ....   Above the boy there is ....

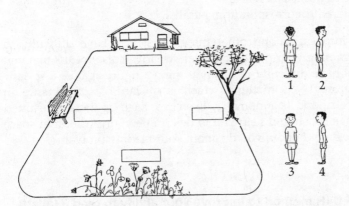

Once again place each of the boys in turn in the middle of the sketch. Complete the following:

Position	Object	Direction in relation to boy
1	The tree	
4		right
2		behind
	The house	in front of
3	The bench	
2	The house	
	The tree	left
4		behind
	The bench	
		left
3		behind
4	The tree	

65

Position in space

This phenomenon or entity is popularly called spatial orientation. We sometimes find that some children do not quite know how to indicate *positions* or do not quite understand words or expressions such as the following:

- near
- far
- three-dimensional
- alternate, corresponding, parallel

This implies, among other things, that you have difficulty in determining your position in space clearly. It also means that you might have difficulty in distinguishing a figure against a specific background – something which is of cardinal importance in geometry and trigonometry. Exercises, such as those illustrated on page 65, should help you to determine your position in space easily. But what can you do about your reading problem?

The PQ4R method to improve your ability to read (Gullatt) mathematics

Tests done with pupils who have difficulty in reading mathematics show very clearly that some pupils have great difficulty in making the connection between words, that others are reluctant to read a paragraph twice ("If I did not understand it the first time, why should I understand it the second time?"), or that some pupils read a number of words which do not appear in the paragraph or sentence.

To handle these and other problems, I suggest the so-called PQ4R method. This is by far the best and easiest way to improve your reading ability. This is how it works:

Step 1: *Read slowly.* The very first step is to read your mathematical problems much more slowly than you have been doing up to now. You see, when you originally learned to read, you started to read stories – and the whole idea was that you should read fast in order to grasp the general idea or even just the *main points*. Reading technical material such as mathematics requires that you should think about it and ponder it much more while reading.

Step 2: *Read again.* The second step is exactly the same as step one: Read every line once more. Now you should apply the PQ4R method:

Preview: Obtaining your first insight into the problem.
Question: Asking questions about the piece.
Read: Read the piece once more.
Reflect: In other words, think about the problem very carefully.
Rewrite: Write the problem down *in your own words.*
Review: Revise everything you have just done.

This process really works with word problems. It boils down to the fact that you:

- slow down your "story reading" speed; and
- find connections while reading.

Step 3: *Learn your technical vocabulary.* I have said it before; I want to repeat it here. Make a list of *all* the technical words you are going to need in your career – depending of course on the standard or year in which you are. Ask your teacher to help you with this list. Know the terms and the symbols well (you will find a list at the back of any mathematics textbook, like the one you presently use in class or for doing homework). Mathematics uses normal words in a special way, with a specific meaning, which differs from the everyday meaning. Make sure that you *know and understand all abbreviations.*

And remember: *Verbal instructions from your teacher should always be cemented in writing.* One day I tried to explain the concept "distributive" to a group of standard 6 pupils. To my amazement one of them had written down the word "astrivoitive" in her book. Note: it was *my fault,* nothing for her to feel bad about. All I want to say is that she should have asked me for the spelling of the word – I am begging you never to make the same mistake.

Step 4: *Adapt your eye movements.* In mathematics (contrary to the normal way of reading from left to right) you read equations with exponents, parentheses, brackets, you look at tables and diagrams – in different ways, moreover. When there are algebraic fractions in equations, pupils usually *read right across the numerator instead of identifying the respective terms as fractions.* Look at the examples:

67

1. $8 - 2 \div (4)^2$

   To simply read from left to right and to work horizontally will be catastrophic!

2. $\dfrac{2x - 5}{3} - \dfrac{4x - 7}{4} = \dfrac{3 - 8x}{5}$

   Do you see what I mean? It makes a big difference when you approach the three terms as three separate fractions – you then eliminate countless negligent mistakes.

Remember: in contrast to the human sciences, we cannot simply read mathematics in order to find the main ideas (as you would do with a story). Here is a short summary of the steps:

A: Try to read *more slowly*, word by word.
B: Read that which you do not understand one more time (and once more, if necessary).
C: Know your technical vocabulary and symbols.
D: Adapt your usual left-to-right eye movements to allow for the variations in mathematics.

Look at the sketch on page 79. We are going to look briefly at that fascinating phenomenon *self-confidence.*

Self-confidence and mathematics

You might wonder why I group self-confidence with language in a discussion of this nature. The answer is simple. Language is based on two things: *feeling* and *knowledge.* In other words: language is the bed of the stream in which your thoughts flow. You cannot think if you do not have control of language. You can only think in words. Therefore, if you have not mastered the language of mathematics, the stream of mathematics cannot flow. Secondly: there is a very close relation between language and feeling. When you are filled with self-confidence, you tend to speak more loudly. When you are lacking in self-confidence, you are inclined to speak more softly and in a husky voice.

What does all of this mean in terms of mathematics? If you are emotionally upset, it has a negative effect on your language, your thoughts and your ability to express yourself. Be sure to remember the name of Carl Rogers. He said that one's self-image is formed and determined by one's contact with others.

Your self-confidence, your perception of yourself, he said, will depend to a great extent on whether *people approve of your behaviour or not.* Therefore, if you discover that other people praise and approve of your achievements in mathematics, it creates self-confidence and the desire to do better. Now do you understand why I encourage your parents to praise and encourage you? Yes, I know:

- Children learn at a very young age that they are "stupid". Failure is very easily *learnt.* It takes on such dimensions that many children find it almost impossible to tackle even an ordinary task, not to mention a sum. It is well known that a pupil with limited self-confidence gives up far more easily than his more self-confident classmates. *Please remember the golden rule in this regard: Failure in one sum, test or paper is simply a learning process. Believe that you can, and you will indeed do better. You will tackle problems with more self-confidence and undeniably have a better chance to succeed.*

- You learn somewhere from an "expert" or informed person that you "have a low IQ". Now you and those close to you believe that if you do well, you are "over-achieving". Nothing can be further from the truth, as you can read on pages 48 and 49. When Pierre was in standard 7 and struggling with mathematics, the head of the mathematics department told his parents: "Pierre has suffered brain damage and he will never pass standard 9, let alone mathematics – he should be in a school for the mentally handicapped." Today he is a qualified mechanical engineer and has also obtained the following degrees: B. Comm., B. Comm. (Hons.), M. Comm. (cum laude) and D. Comm. (his doctorate) – and he holds a top position at one of the best universities.

- You learn from someone that you have a high IQ. If you do not perform accordingly, then you are under-achieving or not working hard enough. Once again, it is a short-sighted and misinformed argument.

To summarize: Research by Purkey has shown beyond doubt that there is a very close link between achievement in mathematics and your self-confidence, your self-image. Remember this golden rule: *It is always better to expect and to think much of oneself, rather than too little.* As Visser puts it: Self-confidence and enjoyment of mathematics are *reliable pointers to performance in mathematics.*

And what about the relationship between music and mathematics?

Mathematics and music

I am going to say very little about the relationship between mathematics and music itself. It has been emphasized repeatedly: Music *is* mathematics. Whether the opposite is also true, differs from one person to the next.

Learning to the sound of music

This selection of compositions was used by researchers to make the minds of students receptive to the foreign languages they wanted to learn. The pieces marked "A" were used where the study material was read in a slow and solemn voice on the cassette. For the pieces marked "B" a normal tone of voice was used. Tapes of the selected pieces can be ordered from the *Superlearning Corp.*, 128, E 56th Street, Fourth floor, New York, NY 10022.

1. (A) J. Haydn, Symphony No 67 in F major and No 69 in B major.
   (B) A. Corelli, Concerti Grossi, Op 4, Nos 10, 11 and 12.
2. (A) J. Haydn, Concerto for Violin and String Orchestra No 1 in C major and No 2 in G major.
   (B) J.S. Bach, Symphony in C major and Symphony in D major; J.C. Bach, Symphony in G minor, Op 6, No 6; W.F. Bach, Symphony in D minor; C.P.E. Bach, Symphony No 1 for String Orchestra.
3. (A) W.A. Mozart, "Haffner" Symphony, "Prague" Symphony, German Dances.
   (B) G.F. Handel, Concerto for Organ and Orchestra; J.S. Bach, The Choral Prelude in A major, The Prelude and Fugue in G minor.
4. (A) W.A. Mozart, Concerto for Violin and Orchestra No 7 in D major, Concerto for Piano and Orchestra No 7 in D major.
   (B) J.S. Bach, Fantasia in G major, Fantasia in C minor, the Trio in D minor, Canonic Variations and Toccata.
5. (A) L. van Beethoven, Concerto for Piano and Orchestra No 5 in E flat major, Op 73.
   (B) A. Vivaldi, Five Concerti for Flute and Chamber Orchestra.

6. (A) L. van Beethoven, Concerto for Violin and Orchestra in D major.
   (B) A. Corelli, Concerti Grossi, Op 6, Nos 3, 5, 8 and 9.
7. (A) P. I. Tchaikovsky, Concerto for Piano and Orchestra No 1 in B flat minor.
   (B) G.F. Handel, Water Music.
8. (A) J. Brahms, Concerto for Violin and Orchestra in D major, Op 77.
   (B) F. Couperin, Le Parnasse et l'Astrée, Sonata in G minor; J.P. Rameau, Pièces de Clavecin, Nos 1 and 5.
9. (A) F. Chopin, Valses.
   (B) G.F. Handel, Concerti Grossi, Op 3, Nos 1, 2, 3 and 5.
10. (A) W.A. Mozart, Concerto for Piano and Orchestra no 18 in B flat major.
    (B) A. Vivaldi, The Four Seasons.

Source: Lozanov, G., *Suggestology and Outlines of Suggestopedy* (Gordon and Breach, 1978)

You can take it from me that noise is not only physically harmful, but it also influences your mathematical learning ability very negatively. Before we go any further, therefore, I want to ask you to make a so-called sound inventory when next you are doing your mathematics at home (which I hope will be very soon). In other words, write down *all* the sounds you can hear – music, talking, traffic, everything. Then write down which of these sounds you find pleasant and which are disturbing. In this way you can sort out which sounds you should try to muffle and which you can try to put to good use.

Music as a barrier

I agree with Halpern that some kinds of music can motivate you, relieve tension, and put you in a better mood for work. (It is a very personal matter and I do not think anybody else can decide what music will affect you positively or negatively.) I also agree with him when he maintains that some kinds of background music can act as a sound barrier against other sounds which may disturb you. Soothing music, which you find pleasant and restful, can act as a barrier against other disturbances and enable you to concentrate more fully on your mathematics.

Music as a starting point for success in mathematics

I want to stress something at this stage: parents should start singing to and with their children from a very early age. It does wonders for the child's development. He learns all kinds of words, learns syntax (the construction of sentences) and generally understands better than his peers who never, or hardly ever, sing. My advice to pupils in this regard is: sing along with each song you hear, and try to remember the words. It really does not matter what it sounds like to you, but remember what Kumon says in this regard: all mathematics should start with a song. The sooner, the better; the more often, the better.

Music and the study of mathematics

My advice to you is the following: Listen to the music of your choice, but not music which agitates you, and *certainly not music which is too loud.* And never listen to music through earphones, I really regard this as an aberration. The reason is that these earphones generate a level of decibels which is quite unacceptable and also concentrate these decibels to such an extent that it can physically damage your ear – without your realizing it. The right kind of music has almost the same effect as yoga or relaxation exercises. Your body relaxes and your brain is clearer. Look at what researchers have found:

Council kicks up a din about earphone noise

Our children are becoming deaf!

A generation of children may become hard of hearing in their teens because of the explosion in modern twentieth century technology: loud music in discos, earphone cassette-recorders like the Walkman, motorbikes, rock concerts and noisy toys.

The Directorate Radiation Control recently carried out tests on six different earphone players. At full volume the average sound level was between 98,8 decibels and 111,3 decibels – as loud as the sound of a jet taking off within 100 m from you! (A sound level of 85 decibels and higher is regarded as harmful to your hearing.)

Mrs Henna Opperman, director of the National Council for the Deaf, is worried about children's ignorance regarding loud music. "The results of loss of hearing are so very far-reaching. It isolates people and affects all relationships."

(Translation from Beeld, 6 June 1991)

What about pop music? I have some good and some bad news: the bad news is that the tempo in heavy pop music (short-short-long-pause) is apparently dangerous because it is the rhythm exactly opposite to your natural heart and pulse beat. Because it disturbs the natural body rhythm, the two hemispheres of your brain lose contact with each other and it lessens your powers of observation, causes stress and lessens your ability to work efficiently.

The good news? It is always better to sit in your own room doing mathematics while you listen to pop music than being somewhere else doing something destructive. Your favourite pop song (and I want to stress that I do not mean what you call "heavy underground": that is really not something you should try to study to) might just be the subconscious motivation you need to get to work. Try to turn the music down very low after a while if you feel that you have to leave your radio or cassette player on while doing mathematics. Music by Rick Astley and Chris Isaak is, moreover, the kind of pop music that gives you something to identify with during stormy times. I am convinced that it can do no harm to do mathematics while listening to it, but that you can put it to positive use. And do sing along at the top of your voice when you have sorted out a problem – relax and turn the music up, until you get down to business in all seriousness again.

To summarize:
- Find the sounds which disturb you and consciously eliminate them.
- Put to use those sounds which help your body and your mind to relax.
- Music which is too loud, of whatever kind, is unacceptable.
- Agitating pop music is good for discos but not for doing mathematics.
- Listen to different kinds of music and then determine which kind you find soothing, inspiring and motivating.

Conclusion

In this chapter we have read about the relation between mathematics, language, your self-confidence and music. In Chapter 8 I am going to give you some (*practical*) guidelines toward becoming a real achiever in mathematics.

CHAPTER EIGHT

Guidelines for becoming an ace at maths

Introduction

The famous French mathematician, René Descartes, once said the following:

We can solve any problem that mankind is ever likely to encounter by following four steps:

1. Reduce the problem to a mathematical one.
2. Express the problem in terms of algebra.
3. Write it down as an equation.
4. Solve the equation.

The last two steps are realistic; I feel up to teaching that to almost anyone. But Steps 1 and 2 are a different kettle of fish ...! Nevertheless, the message is clear: problems are there to be solved – especially mathematical problems.

How do you come to grips with mathematics?

I want to start by drawing your attention to a technical, practical principle which, for one reason or another, always leads to improved performance. *Draw a separate red rectangle* around each important piece of work (such as the examples you copy down in class, the notes you make on mistakes and their corrections, your theorems). This "red rectangle rule" which you repeatedly encounter in this book, *really works*. It is a means of consciously, and also subconsciously, drawing your attention to important rules and principles in mathematics and it makes you absorb these principles more fully than otherwise.

Here are two examples. You will encounter many more in this book:

1.

> *Multiplication in algebra:* $-3x^2y^3 \cdot 4xy^4$
> - $-$ Step 1: Decide what this sign means.
> - -12 Step 2: Multiply the coefficients.
> - $-12x^3y^7$ Step 3: Deal with the letters ONE BY ONE.

2.

> *Golden rule for factorization:*
> *first try to find a common factor.*

Let's look at a few other vital rules in mathematics, rules which will really improve your performance.

Fifteen vital rules you should follow if you want to become an ace at mathematics

A group of my senior students recently had to make a list of the most important rules in mathematics. According to them, a pupil should (in a nutshell) be able to ask himself the following questions every day – and he should be able to reply affirmatively *at any time:*

1. Do you do your homework every day (an hour or more per day, 6 or 7 days a week)?
2. Do you make sure that you understand the work every day (in other words, do you go through it again later in the day, ESPECIALLY JUST BEFORE YOU GO TO SLEEP AT NIGHT)?
3. Do you mark your work every day and do you make a point of correcting wrong answers? Do you follow up tests/examinations, the mistakes you made in them, PROPERLY?
4. Do you take care not to mark incorrect work as correct?
5. Do you know the rules, theorems and formulas in your book *well*?
6. Are you *never* embarrassed to stand up in class and *ask* if you do not understand something?
7. Do you believe that you *can* do mathematics?
8. Can you give reasons for each step in each sum?
9. Is there a notice board in your room with all your test and examination dates and all your test and examination marks?
10. Do you talk to your friends about the work every day, do you discuss mathematical terms and concepts with them?

11. Do you discuss problems of a non-mathematical nature with someone who can help you to solve them?
12. Can you do *fractions* well? (See addendum E.)
13. Do you work *neatly* and in pencil so that you can erase mistakes?
14. Is your work systematic?
15. Do you do more than your teacher expects of you?

The list could certainly be longer but, according to my students, this is a "winning recipe".

To be an ace at mathematics, you have to begin working at it in class, during the lesson.

In your mathematics class

I suggest that every pupil should acquire at least four files in standard 6 or even earlier, in primary school. You should keep these files[1] *neat*, carry them over from year to year and keep them up to date:

- In the first file, you put "things I should know by heart" (like the formula for the surface of a circle, the formula to determine x in the equation $ax^2 + bx + c = 0$). (See addendum A at the back of this book.)
- The second file is for strange words and symbols (see addenda B_1 and B_2). Differentiate between totally unknown words or symbols, and words which have a meaning in mathematics that is different to that in normal language (read page 63 in this connection).
- The third file is your "model examples file". In this you copy model examples of the most important types of examples in mathematics. Ask your teacher to help you distinguish in each case between "unsuitable" examples and model examples – the latter should then be put in your model file (see addendum C).
- The fourth file is very important – in this you note *all* your mistakes. And of course you should indicate every time exactly why you made the mistake and how you are going to set about correcting it. *Always* make sure that you know why

1. Files are better than notebooks because files are expandable and you can add something to them at any stage. Let your father, mother or teacher help you to arrange each file logically and to draw up a proper table of contents.

you made a mistake and, wherever possible, which type of mistake you made (see addendum D).

It is of the utmost importance that you should always, always insert a model example of each new type of sum or demonstration in file C. This means that in matric you can quickly refer to what you learnt in standard 7 about long division by binomials (look at the example on page 117).

The expression "do not sit on your hands" is one a creative pupil used one day when I asked him to tell his friends why he paid such close attention in class. In other words: *take notes*, be active.

A few more hints for the classroom:

- Make sure that you copy *correctly* from the blackboard.
- Do not write down more than you need.
- What appears on the blackboard is more important than what your teacher says.
- *Ask* if you do not understand; write down the answer if you can.
- If your teacher *becomes impatient*: take no notice and ask again. The chances are very, very good that if you do not understand something in class, you will never understand it and you will *not* be able to figure it out on your own.

And now something about your homework.

Homework

To start with: mathematics is like a game. You have to stick to the rules or you are not allowed to take part. The rules are *all-important* – if you break them, then you are "out". If you miss out on something, you will be swamped later on – everything is based on the work you have missed. Everything is founded on what you have already learnt – remember that!

> It is all-important to keep up in mathematics.

Yes, it is true. In history you may fall behind without any problems – in mathematics, as in any foreign language, you will really have a hard time catching up again once you have fallen behind.

Therefore: practice is of the utmost importance. Anyone can understand factorization, or solve simultaneous equations, but it really takes practice to become skilful at this; so skilful that it becomes second nature to you.

Therefore, you just have to do your homework every day and go through the notes you have made in class every day. It is also of the greatest importance to do a sum in your model file from time to time. Write down the reasons for each step slowly and clearly. If you have any doubt as to its correctness, ask your teacher to look at it for you. Furthermore, it is vitally important that you be sure to mark *all* your homework, to make *all* the necessary corrections – and to *keep your mistakes file up to date*. Make a note of *all* the mistakes **you** make, or that you notice other people making. I am referring to mistakes such as the following:

Mistake	*Correction*
1. $-2x(3x-6)$ $= -6x^2 - 12x$	You forgot that: $- \times - = +$
2. $3x = 8$ $x = 5$	You erroneously think that $3x = 3 + x$
3. $\hat{B} = 90°$	In geometry you cannot take anything for granted – it *must* be given.
4. $\sqrt{x^2 + y^2} = x + y$	By going back to an elementary example with figures, e. g. $\sqrt{16 + 9} = \sqrt{25} = 5 \neq 4 + 3$, you will immediately notice your error. See pages 33 and 82.
5. $\sin(x + y)$ $= \sin x + \sin y$	Work out $\sin 60°$; then $\sin 30°$ $+ \sin 30°$ and you will see you have made a mistake

Remember that you are inclined to make certain mistakes repeatedly (as I explained on page 17). If you want to overcome these, I would suggest that you make a summary of them and that you look at them again *in the evening, just before you go to sleep.* Believe me, it works. And that also applies to the theorem or principle which you "just cannot get into your head". If you read through it a few times just before you go to sleep, then your wonderful brain does the rest by itself. Talking of this – your brain is the most wonderful computer which has ever been created. Thank the good Lord for this every day and believe in it, use it, supply it with the necessary mathematical information – and the rest will follow. I am still amazed every day when I see how apparent bunglers suddenly begin to perform well, after years and months of struggle. *Also believe in your own ability.*

Here is a summary of the most important points again:
- Do your homework every day.
- Mark your homework, do all the necessary corrections, never mark incorrect work as correct.
- Keep up the files we have discussed and note all your mistakes in them.
- Read through the things that cause you problems (mistakes, theorems, principles, formulas) just before you go to sleep.
- Keep your work neat.
  Also keep a *large* notice board up to date.

I can be an ace at mathematics!

Notice board

Buy yourself a large notice board straight away and also get yourself a *large* year planner. (You know what I mean: a calendar on which you can see *all the days* of the current year.) Put it up on the wall of your study – and make sure that you plan ahead for the whole year, as far as possible. You should also indicate all test and exam dates on it. If you put important dates and your results on your year planner, you have an overview of the year at a glance. You can put anything important on your notice board. It doesn't matter whether it is the date of your next test or your examination dates or a graphic representation of your mathematics marks during the year, as long as it is something which relates to mathematics. Let your father or mother help you with this – they will do it with great pleasure.

And put up your mathematics objective (an A in matric, studies in medicine, whatever) on the mirror you use every morning to comb your hair, or on your desk – in a place where it will be conspicuous and noticeable.

> *I am going to get an A for mathematics in matric –*
> *I want to study medicine.*

This is an example of something which is particularly motivating, consciously and subconsciously.

Do you know how many times the following question has been put to me: "Please tell me how to learn mathematics"?

How does one learn mathematics?

Let me put it very clearly right from the start. You should not sit down to do mathematics out of frustration when you and your girlfriend have had an argument, when your parents have quarrelled or when you experience any other emotional crisis. *No!* You do not play tennis to become fit (of course you become fit to be able to play tennis), neither should you do mathematics in order to get rid of your problems. You should rather get rid of your problems *in order* to be able to do mathematics in peace. Believe me, the first subject to show up your emotional problems is mathematics. When a pupil comes to me with good marks, except for mathematics, I automatically begin to wonder what is ailing him emotionally. Abstract thought is the first thing to be affected by emotional problems. That is why the insecure chap does not easily venture into a lot of cold figures and unfriendly symbols.

Therefore, if something is bothering you, go and talk to somebody and if you cannot talk to somebody immediately, note it *clearly* on your year planner – where you can see it and where you, as it were, unconsciously say to your brain: you see, I am going to do something about this – just give me the peace of mind (however temporary) now to be able to carry on with my work.

> Wednesday 13 April:
> Rina and I have a date to discuss our quarrel.

Also look closely at the following concentration cycles:

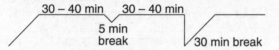

If you try to concentrate for too long without a break, your concentration eventually collapses.

```
      1 – 1½ hours
                      total collapse of
                      concentration
```

By taking a short break regularly (after about half an hour to 45 minutes) and a long break after the second session, you ensure maximum concentration.

Eye strain and headaches deserve a doctor's attention – do not hesitate to seek help. In any case, you should see that your lighting is good and that the distance between your eyes and your book is correct. And try to improve your physical fitness, it certainly helps your concentration. Have you ever come across the following lines before?

> Hear – and forget
> See – and remember
> Do – and understand

* * *

The following two examples illustrate wider thought.

Question 1: Connect all the dots by drawing just four straight lines without lifting your pen.

Question 2: **S B I A X N L A E N T T A E R S**
Remove six letters to find the name of a familiar fruit.
The answers are at the end of this chapter.

I shall now briefly discuss a few steps in the learning process.

> 1. One of the most important insights in this book is the following: If at all possible, look at a simpler form of your problem (or at least look at your problem in a different, wider, way, cf. the figure on page 50). If you can solve a simpler form of the example, then you have learnt enough to solve the more difficult one.

> *How much is x more than y?*
> becomes
> *How much is 12 more than 3?*

> *If I was y years old x years ago, how old am I now?*
> becomes
> *If I was 9 years old 5 years ago, how old am I now?*

82

2. Use what the Americans call the "buddy system" if you wish: study with a friend who is a lot like you *and* who performs well.
3. Use a coloured pen to highlight important formulas, theorems and demonstrations.
4. Learn the facts – know the formulas, theorems, definitions and principles *well*. Remember to revise them within 24 hours. Look at the following two sketches:

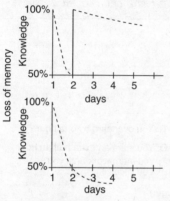

If you learn the work twice (*thoroughly!*) then you remember it *much* longer than when you learn it thoroughly only once.

And at the following:

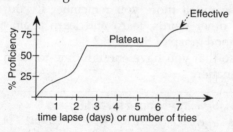

Emenalo and Okpara demonstrate here that it may take as many as seven thorough revisions to really, effectively master the work. And you might take cognizance of the fact that those who really perform well, do indeed work that hard – regardless of what they may tell you. Children often say that they

- did not really learn for the test
- only went through the work very quickly
- really did not learn at all, they did not have the time

and then they perform very well. I want to implore you not to believe them. Take it from me: if you work hard, then you will perform well – and if you do not work hard, then you will not perform well. Whatever your friend's motivation may be to tell you a lie, friends often do that. Never allow that type of remark to make you feel "stupid" if your marks are not as high as theirs, or to undermine your self-confidence – this is extremely important.

> Remember: Mathematics and hard work are synonymous. You are certainly not "stupid" if you have to work hard at mathematics – any maths achiever does well as a result of lots of hard work!!

Look at the following sketch:

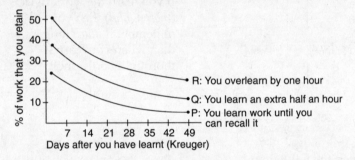

Do you see how much more you remember if you *overlearn* something – in other words, learn and learn again after you already have a good grasp of the work?

5. Plan ahead so that you have enough time to learn before a test or examination.

> Mathematics is not a spectator sport –
> it is largely a memory and figures subject!

6. Do the examples, never just look at them or read through them. Describe what you are doing, make up an example with different figures, this can be of immense value. Here are a few examples of different ways of expressing 88 – 45:
   - subtract 45 from 88
   - from 88 subtract 45
   - take 45 away from 88
   - how much is 88 more than 45?
   - how much is 45 less than 88?
   - what is the difference between 88 and 45?

7. Do a variety of problems, but avoid doing similar problems (which you have already done repeatedly) over and over again *without taking note of the original problem*. Beware of quick-fix solutions.
8. Keep a good balance between learning theory and working out problems. And by the way, the examples in your textbooks were chosen quite circumspectly – work through them and see if you arrive at the same answers.
9. Whenever possible (and especially with longer sums), draw up a scheme. This is how to go about it:
    - Describe the problem.
    - What are the characteristics of the problem?
    - What factual knowledge do you need to solve the problem?
    - Which method are you going to use?
    - Now solve the problem.
10. Don't get left behind – discuss problems with your teacher as soon as possible.
11. Try to lay your hands on as many question papers, textbooks and tests as you can.
12. As far as geometry is concerned, you might follow these guidelines:
    (a) Your demonstration should not only convince yourself – it should convince your examiner in the first place!
    (b) Make a large, clear sketch and keep your captions *readable*. An O which resembles a D, or an A which resembles a D, confuses you and the examiner.
    (c) Try to indicate *all* the data on your sketch. *And use colouring pencils to allow your sketch to "speak to you"*.
    (d) Draw up a plan of action before you start your demonstration. Make sure that you know *exactly* what you want to demonstrate.
    (e) Test your plan before you begin to write. Be *active*, do not just sit and think – think and jot down possibilities until you gain insight. If you really cannot do the demonstration, come back to it at a later stage – your mind will process the matter in the meantime. It is quite possible for you to develop a "blind spot" for the solution if you are too intent on it. A moment's distraction might be just what you need to reach the solution.
    (f) Indicate everything you have already found or demonstrated on your sketch.

85

(g) If you cannot carry on, return to what is given. Make sure that you have interpreted *all* the information correctly.
(h) With regard to each fact, ask yourself the question: How can I use this? Remember that each fact serves a purpose and should be applied somewhere.
(i) If your plan of action does not work at all, make another and try again. Geometry is chiefly a question of try, try and try again.
(j) Look at the sketches below. In sketch 1, which line is the longest? Measure the lines and look for yourself. In sketch 2, which car is the biggest, on the left and on the right? By this I want to illustrate the fact that *geometry sketches can sometimes be extremely confusing – you cannot take anything for granted, you have to be able to prove it.*

13. Apply what you have learnt here – it works.

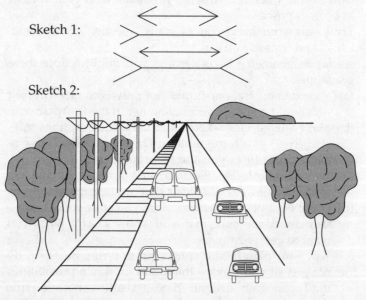

Sketch 1:

Sketch 2:

Extra lessons

I have been giving extra lessons in mathematics for many years now. And if each participant in the class does his part, then it certainly can be effective in making up a backlog, bringing oneself up to date, keeping up to date or even getting ahead in mathematics. But then these rules must be followed:

1. No homework is done during extra lessons. If the extra class turns into a homework period, you are wasting your parents' money.
2. You must be prepared to accept "homework" for the next extra lesson.
3. You must be prepared to write a test now and then during the extra lesson.
4. A time limit should be put on extra lessons, a dual limit:
   (a) Extra lessons should last until your problem has been solved and should be discontinued after that. I have no objection to your asking the presenter of such classes to help you again in three months' time, even if it is only to gain a lead. However, I do want to warn you against over-dependence on him and against the belief that he enables you to perform well. That is certainly not true – *you* perform well after an extra lesson.
   (b) If your performance does not improve substantially within a reasonable time, you should discontinue the lessons.
5. You should always be punctual, go at a regular time and *work neatly* during an extra lesson – it is very important that you should not get into the habit of sometimes working neatly and at other times thinking and working untidily.
6. If, after one or two sessions, you still "only attend extra lessons because my parents tell me to", then I sincerely request that you discontinue the lessons *immediately*.
7. Extra lessons in mathematics are just that – and not a discussion class on your girlfriend, rugby or whatever. (Of course I do not mean by this that you should exclude the possibility of sharing any problem you might have with the presenter if you have developed enough confidence in him.)
8. Extra mathematics remains *extra* mathematics – it cannot replace your normal work and simply means that you are going to work harder at mathematics. You will certainly not find that extra lessons will replace or diminish your input and responsibility.

Kumon method of extra lessons in mathematics

Of course there is *no such thing* as instant results, a miraculous remedy which can be applied in a specific extra class. It is rather a question of every little bit helps – every bit of extra help in mathematics should be welcomed. One method which, in my opinion, adheres to many of the principles set out in this book, is the Kumon method (named after the founder of the classes, Toru Kumon). When he found that his eldest son Takeshi was dropping out in mathematics in 1954, he started to compile work charts in mathematics for him on loose sheets of paper and he found to his amazement that Takeshi improved phenomenally in mathematics. A few principles and guidelines in this method, well worth noting, deserve mention:

- Each pupil first does diagnostic tests to determine *exactly where he is*. It may even be that he first has to learn to draw a straight line! Then only does he begin to work at his level.
- The pupil might have to work through as many as 3 800 work charts in his course. This may sound a lot, but it is not really an unreasonable number. The important point here is that every pupil progresses to a more advanced level of mathematics only after he has mastered the previous level – to the point where he obtains 100% for his work chart.
- Each work chart is subject to a time limit. This ensures that the working speed is acceptably high.
- When a pupil begins to perform well, this reinforces his self-image and self-confidence – and that is the start of the circle: Better performance in mathematics leads to increased willingness to do mathematics and to work harder and eventually leads to a healthier self-image.
- Any pupil, at pre-school or university level, can start this method.
- Repetition, speed and accuracy (three of the foundations for success in mathematics) as well as the fact that a pupil marks and corrects his own work, keeps a record of his own results, receives individualized help, and the accent on *regular work*, make this method something special – and at the same time something which is familiar to all of us.

Of course you can apply most of these principles on your own – but experience has shown that a pupil who has started to struggle only does so in highly exceptional cases.

Preparation for and writing an examination (Botha)

In the first place you should never see a test, assignment or examination as a disciplinary measure. You are fully within your rights, you have the obligation to protest vehemently against anyone who tries to impose this misconception on you. You should refuse point-blank to associate punishment with mathematics, maths homework, a mathematics test or examination. Tests and such like are there exclusively to find out whether you can apply that which you were supposed to have learnt, to new situations. Very often the examples in question papers are quite different from those you encountered before – nevertheless practice in solving problems remains the very best preparation for tests and examinations. Always remember when you are doing preparation to take due note of the mistakes you made before (your mistakes file should come in extremely handy at this stage). Carefully study the tests, homework, assignments, old question papers which were marked and returned to you, to determine what mistakes you are inclined to make. Put extra effort into practising sums like these, and if you have problems working through examples from the textbook, do more and more of them. *And never hesitate to go and ask for help somewhere – help is always available, in the first place from your mathematics teacher and in the second place from a friend with insight.* You need no longer wonder about the sense of an examination. Imagine what it would be like if the doctor who is about to operate on you does not know exactly where and what to cut open because he has never written examinations; or imagine the anaesthetist running the risk of injecting a lethal dose into you because he has never done an examination on the correct doses which he should administer.

A. Preparation for the examination (Botha)

1. Mathematics is not for spectators: *Do* problems, more problems and more. Do them yourself, do not try to memorize examples that other people have worked out.
2. Try to get 7 to 8 hours' sleep before your examination.
3. Do not get up a few hours before your paper and try to "cram". Then you are exhausted before even starting. On the contrary, try to go to bed early the night before.
4. Allocate your time properly by planning your time for every test and paper thoroughly by means of your year planner. And remember: There is no question of "doing badly during the year but learning at the end of the year". It *never* works in mathematics.
5. Plan your revision. See that you allocate your time for three-hour papers properly by doing several three-hour

papers in sessions of three hours, and by asking your teacher to mark them for you so that you can revise your time allocation.

6. Test yourself in writing *and verbally*, discuss mathematics with your friends – it is worthwhile.

7. Concentrate especially on those parts where you know that you experience problems, whether it is geometry, algebra or whatever. Use your old tests and papers at this stage to identify and eliminate problem areas.

8. Prepare yourself emotionally and psychologically for the examination. Believe that you are going to perform well, believe in the mind God gave you – you can rest assured that your Creator always does His part.

9. Do read the contents page of your textbook attentively and make a note of the different sections. Make sure that you can *name* the sections.

10. See that you have a light, nutritious breakfast or lunch before the paper. Beware of too much coffee, tea and cigarettes! Achievement is closely related to physical health. See that you get the necessary exercise and that you are healthy.

11. Collect everything that you need – pen, pencil, eraser, ruler, pocket calculator, handkerchief and headache tablets. See that your calculator has fresh batteries, otherwise the same thing that happened to John may happen to you. Immediately after the start of the examination, he discovered that the batteries of his calculator were flat – with unfortunate results. Do remember your watch.

12. Pay a visit to the cloakroom just before going in to write.

13. Never try to read through summaries or theorems just before writing.

14. Make sure that your clothes fit comfortably.

15. See yourself as a high-jump athlete before an important jump. Be composed and prepare yourself mentally. And of course: Make sure what you are writing – your teacher may sometimes make geometry the first paper and algebra the second! The algebra paper (*usually* the first paper) at the end of matric is made up more or less as follows:
    - Algebra: 70% to 80%
    - Differential calculus: 20% to 30%

The geometry and trigonometry paper (*usually* the second paper) is made up as follows:

- A general question (mixed) covering *any* section: 5% to 10%
- Euclidean geometry: 25% to 30%
- Analytical geometry: 20% to 30%
- Trigonometry: 35% to 45%

Before every test and examination, pupils in lower standards should find out *in the finest detail* what is covered by each paper and what marks are allocated to each subsection.

And in the examination room itself?

B. In the examination room (Botha)

1. Do not talk to friends or fellow-students before you enter the examination room – rather think of something from your past which you found very pleasant. Friends' questions like "Tell me one more time ...", or "What was that formula again?"can tense you up and confuse you.

2. Keep an eye on the time, but do not allow the clock to make you too anxious! Try and work at more or less one minute per mark. Do not get "stuck" on something. Remember: a watch is a mechanical aid – and not a terrifying monster!

3. If it so happens that you cannot do the first or the second question, just relax, say to yourself: I have prepared *thoroughly*, and if I am struggling, then everybody else is probably having a hard time. But do not look at your friends – they might just upset you. Look at anything green that you can find and try again; if it doesn't work, just carry on.

4. Remember that you do not have to start with a specific question. Decide well in advance which parts you want to do first, but do not move to and fro within sections (like series and progressions). Number your questions clearly and *remember: you should not start experimenting in the final examination.*

5. Under no circumstances should you linger on those questions which you are unable to do. Very, very few pupils or students ever get full marks, and if they linger on the sums they cannot do, they may get nil for the paper.

6. If you should "strike a blank", as mentioned on pages 57 to 60, that is no reason to panic. Put down your pen, look at something green, think of something pleasant in the past – or think of the reasons why you are taking mathematics, why you want to perform well, for whose sake you want to achieve good results. Another technique is to go, in your thoughts, to your own special place with your special com-

panion. After a few wonderfully relaxed moments there, your insight should be back with renewed enthusiasm. After a few minutes your "blank" will vanish – believe me.

7. Stay active, read your sums over and over again, draw up schemes and sketches, try to visualize things. *Always see whether a graph might help you reach a solution.*

8. In geometry make large, neat sketches as correctly as possible. Remember that the sketches on the paper can sometimes be misleading, and if something has not been given, you *cannot* assume it. *Always* make sure that the examiner will know which angle you are talking about (B_2 is much easier to find than ABC). Do not just write down anything that comes into your head when you are doing geometry. It is difficult enough for the examiner to correct geometry, without your making it even worse. Always remember that the theorem to be demonstrated in the first part of the sum, will *have* to be applied somewhere in the course of the sum to be resolved.

9. *Never* cross out your answer unless you have *written down* your correction! Otherwise you might just lose marks. At the same time: never give two answers, always cross one out. (Yes, answers in pencil *are* marked.)

10. When doing sums with a calculator, always indicate a few steps, otherwise you might lose all your marks if you make a mistake somewhere.

11. If you have time left, work out those sums of which you are not sure, once again, on a separate piece of paper. You sometimes develop a blind spot for a mistake and can only identify it if you do the sum again *right from the start*.

12. Leave very difficult sums till last, but leave enough space for them. Your brain is like a computer – it will process the matter and *very often* give you the answer when you return to these sums after you have done a few other sums.

13. Remember that in geometry you may use (a) if you want to do (b) and (c), even if you have not been able to demonstrate (a). Simply write, "Presume (a)".

14. See to it that your work is legible and intelligible. Leave enough space between questions.

15. Do not get mad at yourself every time you cannot do a sum. Give yourself a pat on the back every time you can do a sum or can demonstrate a theorem.

16. Relax once you have finished writing. Go for a run, a swim, do anything you enjoy. Reward yourself. You have worked hard, written well and you truly deserve the break – until the next test or examination.

Finally, something about your subconscious.

The role of your subconscious in mathematics achievement

The role of your "subconscious" should never be underestimated. You will notice that I come back to this repeatedly. Yes, believe me, the fact that you continuously make demands on your subconscious, has a kind of hypnotic effect on you. Just as you will cross the barrier in high jump if you believe that you can, so you *will* do better in mathematics if you believe that you can. Use the most fertile breeding ground for subconscious motivation at your disposal. Have you ever suddenly come upon the answer to a mathematical sum while doing something completely different, or had a brain-wave, without being able to explain why? Have you ever woken up in the middle of the night with the answer to a problem, or woken the morning after an examination with the realization that you could have done a problem! When this begins to happen, you are really on the way to success.

Conclusion

In this chapter I have really given away my secrets! You can take my word for it, these guidelines will help you to become an ace at mathematics. That is if you follow my suggestions.

In the next chapter I shall address your teacher.

Answer to Question 1:

Answer to Question 2:
$S̸BƗA X NŁ A ɆN T̸T̸A ɆR S̸$
= B A N A N A

Have another look at the question if you do not understand this answer.

CHAPTER NINE

The part a mathematics teacher can play in making his pupils achievers in mathematics

Introduction

To single out the one teacher who has had the greatest influence on my life, is obviously extremely difficult. But still, without doing anybody an injustice, I want to accord this honour to my own mathematics teacher in matric. He was the one who always inspired me, taught me to believe in my own abilities, gave me self-confidence in mathematics, made me an achiever in the subject. He did this without ever expecting thanks or praise, by presenting his subject with endless kindness, patience and insight. Thank you, Mr van Niekerk, I shall always remember that.

Right at the outset I want to say that I am thoroughly convinced that all children are capable of mastering a substantial quantity of mathematics during their lifetimes. As teachers we have an enormous responsibility in this regard. There is no doubt whatsoever that teachers have the greatest influence on the achievements of our pupils, in mathematics and in their careers. But, and I have to plead guilty too, we are so often overwhelmed that we lose perspective (and heaven knows, it is understandable). As Sarah said when her father asked her why she was doing mathematics so late at night, "I do not understand the work. I wanted to ask the teacher, but he just yells at us …"

First let us discuss the question, "What would a typical achiever in mathematics look like?"

The achiever in mathematics

Suydam and Weaver identified seven characteristics of potential achievers in mathematics:

- the ability to spot similarities;
- the ability to spot differences;
- the ability to spot analogies;
- the ability to visualize and to spot quantitative facts and relationships, and to interpret them;
- the ability to calculate;
- the ability to select procedures and data; and
- the ability to read.

Of course, these are debatable characteristics. The fact of the matter is, however, that we *can* identify enough positive characteristics in each child to be able to use them to motivate him, to encourage him and to stimulate greater expectations in him.

At this stage I want to urge you to avoid stereotyping pupils.

The self-fulfilling prophecy

Do you recognize the following teacher attitudes?

- Pupil A has a high IQ, especially a high non-verbal IQ: this means that I expect him to perform well in mathematics. If he does not do so, then he is lazy.
- Pupil B has a low IQ; if he does well, then he is an over-achiever.
- Boys and girls are treated differently in class – boys are seen as the real achievers.
- Pupil C has such an enormous socio-economic backlog that I just cannot help the poor child.

These attitudes imply that we are in fact making a prophecy about mathematics – and then we try, consciously or subconsciously, to fulfil this prophecy. It really is a great pity.

And what do authoritative authors say about the mathematics teachers who are likely to achieve success?

The ideal mathematics teacher

According to those who are knowledgeable, a mathematics teacher should possess the following characteristics:

- a thorough knowledge of mathematics, beyond the level which he is to teach;
- a sincere and genuine interest in the subject;
- appreciation for the language of mathematics;
- an awareness of the applications of mathematics and the connection between mathematics and other subjects;
- an awareness of the wide variety of reference works on mathematics, available to promote effective teaching (this includes books, diagrams, films, apparatus and everyday materials in our immediate environment);
- a broad knowledge of the history of mathematics;
- good knowledge of the typical problems with which children struggle in mathematics;
- an appetite for mathematical puzzles and access to a wide variety of them;

- thorough knowledge of the particular problems of gifted pupils and slow learners;
- a mathematical sense of humour. Repeatedly asking sarcastically, "Professor, what is your father's answer this time?" is certainly not humorous. On the contrary, it damages a child's self-image in mathematics.

A mathematics teacher can be forgiven if he (sometimes) teaches badly, we all have our "off" days. However, I do not forgive him if he teaches mathematics incorrectly – "Log –6 + log –3 = log 18" or "if you calculate $\frac{22}{7}$ then you find the value of pi!" Or when he neglects to tell his pupils repeatedly that mathematics is also a *learning subject*. Incidentally, where does the misplaced habit of grouping mathematics and a "learning subject" together on the same day in examinations originate? That only reinforces the misconception that one does not have to learn for mathematics.

I cannot forgive a teacher who makes a mistake and lacks the courage to admit it, or who cannot say on occasion, "Oh dear, I really do not know – but let me find out and I will come back to you." What is wrong with that? In the words of the *Guardian*: "It is the maths teacher that matters most in the end." And if I may quote Kipling: "We have 40 000 000 reasons for (poor teaching), but not one single excuse!"

Could a teacher be too "clever" to explain the subject? Possibly. But it does not have to be the case. The alternative is not to work with children.

A remark to conclude this section: Have you ever noticed that the facts in nearly *all* textbooks are outdated by about 10 years? (If one loaf of bread costs 5c … I read in a primary school mathematics textbook; in a standard 6 book I came across: if 12 apples cost 85c …) I really want to urge you: Teach real-life mathematics. But more about this on pages 106 to 109.

Now we have looked at the different characteristics of teachers. What do the experts say about their teaching?

Effective teaching strategies in the mathematics class

I do not want to use this book to make a stirring call for technical education, but … as Welch puts it: "Local factory managers may be impressed by job applicants who can quote Shakespeare, but will they hire them? … Mathematics … should be taught in more practical – but not easier – ways. All vocational education courses should become ... more relevant to students' future needs."

Here are a few general principles which relate to the effective teaching of mathematics:

1. All children are willing to learn *something* – teachers have to determine their capacity to learn.
2. Success is of the utmost importance. Immediate feedback is just as important.
3. Self-concept influences achievement and vice versa. Children should have worth – in their own eyes and in yours.
4. Practice is vital.
5. A teacher should be aware of various teaching strategies – when one fails, he should have another to fall back on.
6. Children with problems in mathematics may well work slowly – more time is often required to cater for their needs.
7. Children's mistakes have to be analyzed very, very thoroughly. There is usually a reason for mistakes, although children do make negligent mistakes from time to time. Teachers should look for patterns in the children's mistakes, discuss their thought processes with them and set the erroneous concepts right.
8. Each child learns in his own, individual way.
9. Children learn best when they are highly motivated.
10. Children convey their knowledge more easily when they are solving *real-life* problems.
11. Each child should be informed individually regarding the extent to which he is performing well or badly within the context of his own objectives.
12. Children learn best when they participate *actively* in class.
13. Children learn best when they can communicate – especially when they can discuss their insights with their friends.
14. Children should soon learn that what is important in mathematics is not the right answer, but *learning*.
15. Children learn best when feelings of frustration, inadequacy and fear are reduced to a minimum. Do remember that negative attitudes are often learnt at home. When the levels and speed of learning are set in such a way that the children can taste success, they feel safer with their teachers, friends and mathematics.

How does a teacher motivate a pupil in mathematics?

Motivating mathematics pupils

First of all, there are a few concrete measures which can be taken:

- putting up a roll of honour in the mathematics class;
- inviting pupils to become members of a mathematics club;
- individual interviews with counsellors and maths teachers;
- letters to parents to praise achievements;
- mathematics competitions;
- the availability of a wide range of books on mathematical principles and instruments.

The very best instrument still remains the mathematics teacher's personal example and enthusiasm, his love of mankind and his compassion. As long as his pupils have positive feelings about him, they will continue to perform well in his subject.

I often refer to Piaget and the stages he identifies in the child's mathematical development.

Knowledge of the child's environment and heredity

We have neither the time nor the space to go into this. However, I refer every teacher to people like Piaget, Copeland, Skemp and Gagné for further information.

I just want to stress this one aspect very strongly. At about the age of 11 or 12, according to Piaget, the child passes from the stage of concrete thought to that of formal thought processes. This goes hand in hand with the child's emotional and physical development (when he reaches puberty) – and numerous authors point out that pupils undergo a critical change in attitude towards mathematics just at this point, when they undergo crucial intellectual development. Why do I bring the subject up here? Because you should, for instance, take note of the fact that the child becomes a *group person* right at the beginning of his junior secondary school phase, and this has important implications for your teaching methods.

At this stage I want to comment on the so-called gifted child. The Americans refer to *main-streaming* and by this they mean the process of placing gifted children in "ordinary" classes. Suffice it to say that I am a strong proponent of this, rather than taking such a child out of that particular situation.

Handling highly gifted pupils in your class

Teachers often wonder whether or not it is right to seat the more gifted pupils (or those who perform well) apart from the rest of the class. I shared my views with Mr Chaka van der Merwe, headmaster of Die Wilgers High School, and he agrees with me wholeheartedly. This really says something, as he is generally regarded as one of the most successful mathematics teachers ever! Here are some guidelines in this respect:

In every class there should be a clear distinction between three groups, higher grade, standard grade and lower grade. You may arrange the class in such a way that these groups can be easily identified. But do not let your more successful pupils in the higher grade sit apart from their group. And it is not necessary to give them reams of extra sums to do, either. Rather stimulate them in a less conspicuous way. Ask work-related questions which are "different" and more difficult than the usual questions in the book. Accommodate them in the appropriate group, but make them *think* and *puzzle*. For example, towards the beginning of standard 9, when you are doing the instruction for the parabola, and you explain the equation $y = ax^2 + bx + c$, you can ask these pupils to go and find out why $a < 0$ gives a maximum and $a > 0$ gives a minimum value. See whether you can guide them to discover differentiation by themselves. Be creative – and do not "spare" these pupils. Tell them when you are not satisfied with 83% for a paper. It may seem strange that I, of all people, should ask you to put pressure on a pupil, but it is a fact that life is not going to spare these pupils when they grow up. They are the leaders of tomorrow and very often they function best under pressure.

In the penultimate section of this chapter I shall briefly discuss the analysis of children's mistakes.

Analyzing children's mistakes in mathematics

In an end-of-term test two pupils received 0 out of 2 for the following sum: $2x - 3(x - 4) = 5$.

Pupil A's answer was the following: $2x - 3x - 4 = 5$, $x = 4$. In this case, I agree with the nil. Pupil B, on the contrary, answered as follows: $2x - 3x - 12 = 5$, $-1x = 17$, $x = -17$. Actually, very good, except for one mistake. I want to make bold

100

to say that we often do not take the trouble to really give suitable recognition to achievement – even if it is partial achievement.

To my mind we should always approach a mistake in mathematics as the result of a failure – not the cause. Underlying the majority of problems, and concealed behind them, is another, more important problem. It depends on us whether we are going to take the time and the trouble to search for it.

I want to urge you to correct your pupils' work, make notes of common and other errors and discuss them with the whole class, on a very regular basis and as often as time allows. This is of vital importance. When a teacher marks the following correct, and writes "Good" next to it, it is almost as bad as when the teacher stops at a pupil's desk, draws a line through a wrong sum and writes next to it, *"You idiot!"* (This actually happened.)

Demonstration:
$$x^2 - z^2 = (x - z)(x + z)$$
$$x^2 - y^2 = x^2 - zx + xz + z^2$$
$$x^2 - z^2 = x^2 - z^2$$
$$0 = 0$$

And as far as our papers are concerned, it is quite unforgivable when a pupil in one of our posh schools has such a hard time trying to read the illegible words, figures and letters that she comes nowhere near finishing the paper at all. Always bear in mind that legible typing and printing (if you really have to write, it *must* be very clearly legible) should at all times be the most basic requirement of a paper.

There is nothing wrong with pupils marking each other's work from time to time – there simply is not enough time to mark each pupil's work every day – but you really must check these answers at some or other stage.

With regard to pupils' mistakes, I must emphasise that mistakes can be the trampoline to success for any child. Putting mistakes to good use can provide a foundation for growth. However, if you give a child the impression that failure is devastating, bad and unforgivable, he develops all kinds of undesirable reactions or stratagems to avoid failure, like copying sums from a friend. In a class situation, when you ask a question and tell the pupils to put up their hands when they know the answer, one may put up his hand and then mutter incoherently if you should chance to ask him for his answer; another may suddenly demonstrate complete disinterest; another may look very interested in something else; and yet another may begin to answer and then suddenly hesitate. This is simply inadmissible.

To summarize this section on children and their mistakes, I should like to pass the following on to you. I beg of you, *always* remember the following basic rules concerning mathematical mistakes made by your pupils:

1. You should at all times have sympathy, no, rather *empathy*, with children and their mistakes. Children do not make mistakes simply because they are "stupid". Their mistakes are meaningful, rational attempts to handle mathematics. These mistakes originate in what they have learnt. Psychologically speaking, and from the pupil's perspective, these mistakes make sense (Olivier). Always take this into account before you become too seriously upset about mistakes.

2. Traditionally, the university always blames the high schools for students' poor performance; the senior secondary section blames the junior secondary section, the junior secondary section blames the senior primary section; and so on ... But that is not the point. Everybody should rather do their best and leave this quarrel to those who can best deal with it.

3. I have repeatedly pointed out that mathematics is a subject which inexorably builds on previous knowledge. You will agree that:

   - to learn new work correctly, previous work must have been learnt correctly;
   - by the same token, learning "incorrect" mathematics is sometimes founded on the incorrect mathematics which was learnt previously;
   - learning "incorrect" mathematics is, however, mostly the result of learning *correct* mathematics. Each mistake made by pupils, has its origin in a principle which is correct – but which is applied or interpreted incorrectly, or applied partially correctly. For example, it is perfectly correct to write 2,5 as $2\frac{1}{2}$. However, when you want to say that you multiply 2 by $\frac{1}{2}$, then you *must* write $2 \times \frac{1}{2}$, $2 . \frac{1}{2}$ or $2(\frac{1}{2})$. It is also true that $2(3 + 4) = 2 \times 3 + 2 \times 4$, *but* $\sin (20° + 30°)$ is certainly not $\sin 20° + \sin 30°$!

4. The origin of misconceptions is often (mainly) an over-generalization or incorrect generalization of previous knowl-

edge – to the point where it is no longer valid in a new and different field.

5. Once a pupil has learnt an erroneous scheme or thought pattern, this scheme or pattern resists change fiercely.

6. Children often shrink from learning totally new schemes. They would much rather try to fit their new ideas into "old" schemes, to the point of distorting them in order to fit them into familiar schemes.

7. New knowledge, facts, principles, theorems cannot simply be taught to children as something totally different. You should much rather see mistakes and the reasons for mistakes (misconceptions) as a natural consequence of the children's efforts to expand their own knowledge of mathematics. Mistakes can *never* be avoided. No less can they be regarded as monsters which should be eradicated completely and at all costs. That is enough to make the child lose his self-confidence completely. You should much rather see mistakes as part of the learning process. You should create an atmosphere in class in which there is room for mistakes and try to exploit mistakes as an opportunity to promote learning. Therefore, simply pointing out mistakes and misconceptions to the children is of either very little or no help and does not serve the purpose. You should repeatedly help your pupils to *make the connection between new knowledge and previous knowledge*. Always link up mistakes with previous mathematical knowledge, in other words, be positive about mistakes.

8. I know that one usually wants to help as many pupils as possible in each period. However, taking your pen and simply doing a pupil's work for him, really is a negative action – try not to do it. One of the best mathematics instructors I ever had at university, once had the following effect on one of my fellow students: He threw his hands in the air and said, "It's no use asking him for a solution, he acts as if he is the one with problems and keeps asking me questions." I must add that once he had done with you, you knew where your problem lay.

9. Years ago, two teachers marked the same matric pupil's second (geometry!) paper, quite independently of each other. One gave him 28%, the other 94%. In other words, even where personal prejudice does not play a role, we sometimes make *big* mistakes in mathematics. You and I are *human*, we

will make mistakes. Let us admit that, to our pupils as well, and try to keep these mistakes to a minimum.

To conclude this section, it is a good idea to take stock from time to time, to determine your pupils' attitude towards you and your subject.

Opinion polls in the mathematics class

There are mainly three kinds of (preferably anonymous) questionnaires which you can use from time to time to determine the pupils' attitude towards the subject. In the first place you can ask them to make a list of all their school subjects and to label the subject they like most number 1, the second best number 2 and so on.

In the second place you can draw up a short, descriptive questionnaire, something like the following:

MATHEMATICS IS:								
Fun	7	6	5	4	3	2	1	0	*A punishment*
Meaningful	7	6	5	4	3	2	1	0	*Useless*
Important	7	6	5	4	3	2	1	0	*Unimportant*
Necessary	7	6	5	4	3	2	1	0	*Unnecessary*

In the third place you can compile the following type of questionnaire:

	Always	*Some-times*	*Never*
I like mathematics			
I like doing mathematics			
I want to carry on with mathematics			
I do my homework in time			
I perform well in mathematics			
I use mathematics outside the school context			

You certainly do not need a doctorate in statistics to be able to draw meaningful conclusions from these questionnaires. In this way you can determine what the average (or the individual) pupil's attitude is towards the subject. And you can begin to look for reasons and causes for problems. Indeed, through the eyes of others we see ourselves as we truly are.

Conclusion

In this chapter I have pointed out that you are the child's most important companion in mathematics. You can truly make or break him and you can make the difference between success and failure in the mathematics class. This surely places a particularly high moral responsibility on the shoulders of every mathematics teacher.

In the last chapter you will read more about the relation between culture and mathematics, and we will discuss the question whether mathematics problems are learning problems or achievement problems.

CHAPTER TEN

Mathematics and culture. Are mathematics problems learning problems or achievement problems?

Introduction

Children come from various home environments and have different backgrounds. There are children from affluent homes and children from deprived homes; children differ as to their ethnic and cultural backgrounds; and motivation differs from one culture to the other, as does children's interest and the value parents place on learning.

Children from a more stimulating environment have a wealth of experience and often learn more easily. In the same way, it sometimes happens that children from deprived homes (non-stimulatory environments) struggle and learn more slowly as a result of limited experience.

In this changing country these are aspects which should be viewed in all seriousness by mathematics teachers.

Should mathematics be taught multiculturally?

It is generally accepted that mathematics is an international language and is therefore culture-free. This is very far from the truth – mathematics has many culture-related nuances. One might indeed say that mathematics is of less value in the creation of a multicultural society than, say, religious instruction – but that would be a misconception about the inherent strength and character of the subject.

What presumptions are made about culture in teaching mathematics? I am going to name a few of the more important ones:

106

1. "White middle-class children do better than their black counterparts because of the intrinsic nature of the subject." Is it not true that we teach white middle-class mathematics, to black children too, and that they are excluded from the subject because their culture is different? Or, even closer to the truth, do we as teachers perhaps make assumptions about the mathematical ability of white middle-class children and convert this into a self-fulfilling prophecy through our teaching?

2. "The content of the mathematics textbook is of secondary importance – what is important, is the mathematical quality of the content." It is, however, tragically true that certain lifestyles are over-represented in the practical problems in textbooks. Examples are, planning a seaside holiday or calculating share prices. An even greater problem is the medium of instruction: even where instruction is given in English, the language is often inappropriate for those who speak a different brand of English, who do not speak standard English, or who normally speak an African language.

The fact that the names and backgrounds which occur in mathematics textbooks certainly do not represent the black working class, that the illustrations are of white pupils, that the activities are those of whites, can very easily give black pupils the impression that mathematics is not for them, not about them and that they do not belong there. The fact that Asian and African mathematics played an important part in the development of the subject is often ignored, while the Greek and European contributions are elevated to the only contributions. We might do well to note that the word algebra, for instance, is derived from the first word of the book *Al-jabar w'al Mugabulu* which was written by an Arab mathematician as far back as 820 years before Christ. The contribution of African and Indian mathematicians to the development of the subject (like the invention of the number 0, the principle of place values, our number system, algebra) should also not be underestimated.

What is the reason for these and other misconceptions? Teacher training courses that are inadequate? The influence of the mass media? The classroom culture which has developed in our schools over the years? Probably a combination of these and other factors.

Let me make it quite clear that skin-colour is not a prerequisite for being an ace at mathematics. It only means that everyone should have the opportunity to be able to develop to his maximum potential.

The multicultural role of mathematics

How can mathematics help the child to understand the world around him more fully? Look at the following two examples and compare the impact of the two sets of remarks.

1. Hunger is a problem	40 000 children die of hunger/infection every day
2. Women are discriminated against	Women • represent 50% of the world population • work 67% of man-hours in in the world • earn 10% of the world's income • own 1% of property worldwide

What am I trying to say? The creative teacher, who is worth his salt, can help every pupil in his class to make sense of the world around him, to order his world, to really become a winner. Read on and see if you agree with me.

These days we teach mathematics, at school and at university level, in such a way that we place heavy emphasis on concentration, self-discipline, accuracy, conforming to rules, silence, perseverance, precision and sophisticated language usage. The question immediately arises whether we should not teach mathematics in such a way that we also promote creativity, group cohesion, intuition, imagination, the ability to express oneself and greater freedom in doing so. Why? In the early *and* later stages of teaching mathematics these characteristics and attitudes are an advantage in mathematics – ask any expert who is worth his salt.

To my mind, the way in which mathematics is taught at present seriously discriminates against certain characteristics that are mainly culturally determined. Certain Muslim communities, those with an authoritarian, disciplinary and group-dependent character, strongly relate to this method of teaching (as does the traditional Afrikaner community); while the method of teaching mathematics which is propagated here, is better suited to the more "liberal" and relaxed attitude to life which is typically English, as well as to the natural exuberance of traditional black peoples.

Is mathematics only accessible to certain types of people? Of course not. But then these factors have to be discounted timeously in our teaching of mathematics.

The role of visualizing in mathematics deserves the same measure of attention.

Visualizing in multi-cultural classrooms

(See also pages 44 to 47.)

The most authoritative research has shown that visualizing (representing data by means of a sketch) can be a powerful aid in any mathematics class. In all cultures, and apparently even more so in non-Western cultures, the schematic representation of mathematics is an exceptional aid in teaching the subject. In fact, the idea of "mental pictures" is gaining ground everywhere – twelve-year-old Neil D. expressed it like this: "I use a picture to find out what's going on, to get the idea."

It is of the utmost importance that teachers who have children from different cultures in their classes should realize that the need for "pictures" or visualizing is of even greater importance when teaching takes place in a medium other than the child's mother tongue. In fact, it is true of all pupils that visualizing can facilitate insight into mathematics; insight which is handicapped by a lack of fluency in the medium of instruction.

But there is also another, more subtle, type of language problem which teachers in a multi-cultural classroom have to take into account.

Problems relating to pupils and curricula

I would like to illustrate the problem: When research was done into the reasons for particular problems with mathematics in Botswana, the following came to light: the British mathematics curriculum of the time made virtually no concession to local culture, for example a preference for *small* numbers, a taboo on indicating the exact number of cattle, a rich vocabulary for individual heads of cattle and the immense importance attached to the here and now. The problem was solved by adapting the curriculum to fit in methodologically with the learners' natural way of learning.

What do I want to demonstrate in this section? That as mathematics teacher you will have to go back again and again to the individual learner if you want to make a successful pupil of him. And that it is possible, but that it will require a sustained effort and sacrifice on your part.

In conclusion, is a problem in mathematics primarily a learning or an achievement problem? As we have seen above, it is impossible to pass judgement unless one goes back to the individual learner.

Failure in mathematics (Gannon and Ginsburg)

Keep Piaget's words in mind: "Failure in mathematics is a *symptom and not an ailment*." We simply have to look for the reasons. Let us first focus on learning problems.

Learning problems

In some cases, failure in mathematics is an undeniable indication of learning problems. There are various ways in which the learning process can be disrupted.

1. Teaching deficiencies. Some examples:
   - Some teachers (especially at primary school level) are easily intimidated by mathematics, are scared of it, or are simply ignorant of it.
   - At some colleges teacher training includes very little mathematics and very little is learnt about children's thought processes.
   - Teachers are ignorant of alternative methods of presentation, while flexibility, different methods of explanation, and multiple means of expression are of the greatest importance in mathematics.
2. Emotional problems. In this case, the teaching is adequate but emotional problems hamper the pupil's progress.
3. Problems of style. For example, if a pupil prefers to learn by self-discovery, a teacher who stresses mere memorization and acceptance, will create problems for such a child.
4. "Bugs". Brown and Burton refer to these as consistent, repeated and logical strategies which systematically lead to calculation mistakes. These bugs are usually a simple distortion of certain algorithms in mathematics:

$$\begin{array}{r} 34 \\ +21 \\ \hline 73 \end{array}$$

becomes:
Can you see where the problem lies?

The pupil simply added up the top two figures and then the two below. Of course the answer is incorrect, *but* the child could add correctly, only the direction was wrong. Therefore: "bugs" of this kind do not represent a fundamental deficiency

and can be corrected, provided that someone takes the trouble to uncover them and analyze them.
5. In the last instance, there really are a few children who simply cannot do mathematics. Apart from mentally retarded children, these children constitute a very small minority.

But performance problems are a more general cause of problems in mathematics.

Performance problems

These are unlike learning problems. The pupil experiences no problems in learning the work; something merely influences his rendition of what he has learnt and consequently his performance in mathematics. A pupil with a performance problem is a pupil who understands mathematics in one context but who is not able to reproduce it in a different context. Boredom, fatigue, cognitive style (for example the overly perfectionist child who wants to do everything right and then has a problem with time) are all reasons for this phenomenon.

Therefore, always remember that one cause can have a different effect on different children. Where emotional problems prevent one child from learning, they prevent another from performing.

Implications for the teaching of mathematics

Because problems have different causes, the treatment of these problems will differ. We can represent *learning problems* as follows:

Wall

Pupil A   Mathematics content

In other words, A cannot come to grips with mathematics. He cannot memorize it, he is forever unable to do his homework – in short, he struggles to learn mathematics. It can be an emotional problem (as we have just read) or something else which prevents him from learning the work – he struggles to *learn* mathematics! Teachers and parents can work with the child to remove this wall. Once it has been removed, the pupil can continue with the subject unhindered.

Performance problems can be represented as follows:

Wall

Pupil B   **Performance in mathematics**

Contrary to the first case, B does not struggle to learn mathematics, to understand mathematics or to do his mathematics homework. He *knows* his work, as we have seen, but his *performance is poorer than it should be, if one takes his knowledge of mathematics into account.*

The solution to this kind of problem lies in teaching the pupil how to apply his knowledge, in strengthening his self-confidence and in concerning ourselves less with the content than with the pupil's way of handling it.

Conclusion

It is my firm conviction that all children have the ability to do well in mathematics at school. By far the majority of problems with mathematics originate in social, emotional, personality or educational problems. By really making an effort to disentangle these problems and to smooth them out, we can best answer to our God-given vocation, which is to help others to become whole again; to become and to remain achievers in mathematics!

ADDENDA

Addendum A: List of examples of principles, rules and demonstrations which you must always know by heart

1. $a^m \cdot a^n = a^{m+n}$

2. $a(b + c) = ab + ac$

3. The angle subtended by the arc of a circle at the centre is equal to twice the angle subtended by the arc at the circumference. (*Make a sketch wherever you can!*)

4. $\dfrac{dx^n}{dx} = nx^{n-1}$

5. a. The turning point of a graph with the equation
   $$y = ax^2 + bx + c \text{ is: } \left(\frac{-b}{2a} \; ; \; \frac{-\Delta}{4a} \right)$$
   b. Another way of calculating a turning point: Simply substitute $x = \dfrac{-b}{2a}$ in the original equation.
   c. Another way in which to determine x in the turning point: Differentiate the original equation, let the derivative be $= 0$ and solve for x.

6. $S = \dfrac{a(1 - r^n)}{(1 - r)}$ in the case of a geometrical series
   (see list of terms for explanation);
   $S = \dfrac{a}{1 - r}$ if r is a fraction and if n tends towards infinity (i.e. if it becomes very large).

Try, wherever possible, to establish links between different sections of the work – as indicated in (5). Keep this list complete and up to date – make very sure that you have also noted all important demonstrations in it!

Addendum B$_1$: Mathematical symbols

Symbol	Explanation
1. Σ	The sum of
2. $\lvert -2 \rvert$	the "absolute" value of -2 is $+2$
3. $\parallel$	parallel
4. m	gradient
5. $\dfrac{\Delta y}{\Delta x} = m$	$\dfrac{\text{change in } y}{\text{change in } x} = \text{gradient}$
6. $2 < 5$	2 is smaller than 5
7. $\triangle ABC \parallel\!\!\parallel \triangle DEF$	$\triangle ABC$ is similar to $\triangle DEF$
8. $(2;\ 3)$	numerical pair: $x = 2;\ y = 3$
9. Δ	discriminant $(= b^2 - 4ac)$
10. x	any number

1. Of course my list is not even nearly comprehensive. However, you must keep up, expand and retain such a list!
2. Make very sure that you are able not only to recognize the symbols and understand their meaning, but that you also regularly say out loud what they are in "English".

Addendum B₂: Mathematical terms

Keep up a list of all the terms that may be used in mathematics – and note their specific meaning. Make sure that you understand all the terms.

Term	Example	Explanation
1. Expression	$2x - 3$	one or more terms – not an equation
2. Identity	$3x + x + 1$ $= 4x + 1$	we call it an "identity" when both sides are identical, although they may be written in different ways
3. Roots	$2x = 4$	the value of x for which the equation $x = 2$ is true
4. Altitude		AD is the altitude; that is, a perpendicular line from the top angle to the base.
5. Equation in x of the first degree	$2x + 3$ $= x - 6$	in other words, x^1 is the highest power of x that occurs
6. $\frac{dy}{dx}$ or the derivative of y in relation to x	$\frac{dx^2}{dx} = 2x$	equation for the gradient of the tangent on y at that point
7. Concentric	$\odot A$ and $\odot B$ are concentric	the two circles have a common centre; in other words A = B
8. Discriminant	$\Delta = b^2 - 4ac$	deduced from $y = ax^2 + bx + c$; determines the roots and their nature

Actually you should start to keep this list in grade 1, keep it up to date and always retain it. It is especially important that you should use these terms repeatedly and that you say them out loud. And again: try to learn AND TO USE as many terms as possible in English with a view to further study.

Addendum C: Extract from model example file

You do long division in algebra for the first time in standard 7. A model example might look something like this:

$$2x - 5 \overline{\smash{\big)}\, \begin{array}{l} x^2 + 7x \\ 2x^3 + 9x^2 - 33x - 5 \\ \underline{2x^3 - 5x^2} \\ \, 14x^2 - 33x \\ \, \underline{14x^2 - 35x} \\ \, 2x - 5 \\ \, \underline{2x - 5} \end{array}}$$

1. Divide: $\dfrac{2x^3}{2x} = x^2$
2. Multiply:
    $x^2(2x - 5) = 2x^3 - 5x^2$
3. Subtract
4. Bring down
    Repeat the process!

When you come across this again in standard 9, your model example file will be available immediately.

Let this example suffice. Just remember:
(Place these model examples within a red rectangle – this will make the work stand out immediately.)

1. Keep a notebook for the *work done each year* – and take it with you to university, technikon or technical college.
2. Make sure that you have copied down your model example *correctly*.
3. Make sure that you understand your model examples thoroughly.
4. *Use* this book as often as possible – for reference purposes, for revision, to create similar examples in your scribbler. But *please*, always keep it *tidy*!
5. Remember to draw up an alphabetical table of contents for your file.
6. *All* the important facts that you might forget, should be in there.

Believe me – if the advice of top students means anything to you – this system works!

Addendum D: Errors file

The following examples of errors and their corrections should give you an idea of exactly what I mean by this. Note that you:

1. should preferably indicate the error in red ink;
2. should study these errors and do similar sums before *each* test or examination and also, as often as possible, before falling asleep at night;
3. may well distinguish between the mistakes you made in class (let's say you put a red cross next to these sums) and those made by others (let's say with a black cross);
4. should always *make sure* that you know exactly why a certain error was made and that you write down the reason in a way that is *clear to you*. Above all, use a test or examination paper regularly to look for sums of this nature and concentrate on them;
5. should *always* have this file with you in class – it should be an intrinsic part of the contents of your satchel, like your stationery;
6. should know that saying to yourself: "I shall write it down later", is a recipe for failure – you should always note errors and their corrections *immediately*;
7. *must always concentrate on the type of sums in which you have previously committed errors, rather than on those which you always get right. This is of the utmost importance since there is little purpose in doing sums which you "overlearnt" long ago. In order to obtain higher marks, you simply have to start ironing out your mistakes. It works like a snowball: the more you succeed in ironing out problems, the higher your marks will become and the better you will be able to iron out other problems;*
8. should remember your "red rectangle rule": everything you should know, all model examples and *examples of mistakes* ought to be in a neat red rectangle – it helps to draw your attention, to make you concentrate and also to give you the subconscious impression that this work is a *must*;
9. remember to draw up an alphabetical table of contents for your file!

Standard 5

$$\begin{array}{r} 701 \\ 7\overline{)5114} \end{array}$$ Correction: $\begin{array}{r} 730 \text{ remainder } 4 \\ 7\overline{)5114} \end{array}$	$51 \div 7 = 7$; the 2 that remains must not be added to the "ones" BUT to the "tens"
$$\begin{array}{r} 79 \\ + 32 \\ \hline 165 \end{array}$$ Correction: $\begin{array}{r} 79 \\ + 32 \\ \hline 111 \end{array}$	We do not add the numbers *next to each other*, but those *underneath each other*
$$\begin{array}{r} 464 \\ - 238 \\ \hline 234 \end{array}$$ Correction: $\begin{array}{r} 464 \\ - 238 \\ \hline 226 \end{array}$	In subtraction I cannot ignore the minus sign and simply subtract each smaller number from the larger
$$\begin{array}{r} 84 \\ - 28 \\ \hline 66 \end{array}$$ Correction: $\begin{array}{r} 84 \\ - 28 \\ \hline 56 \end{array}$	In subtraction you must always remember when a 10 has been "borrowed"

Standard 6

$8 - 2 \times 3 = 18$ Correction: $\begin{aligned} 8 &- 2 \times 3 \\ &= 8 - 6 = 2 \end{aligned}$	In arithmetic we multiply before we subtract ! Remember the sequence: (); of; $\div$ and $\times$; $-$ and $+$

$$\frac{14\,x^{14}}{2\,x^2} = 7\,x^7$$	
Correction: $\dfrac{14\,x^{14}}{2\,x^2} = 7\,x^{14-2}$ $\phantom{\text{Correction: } \dfrac{14\,x^{14}}{2\,x^2}} = 7\,x^{12}$	We divide the numerical coefficient of the denominator *but* we subtract the coefficient of the denominator from that of the numerator
$$4(2a + 3b) = 20ab$$ Correction: $\quad 4(2a + 3b)$ $\phantom{\text{Correction: } \quad} = 8a + 12b$	We cannot add $2a$ and $3b$; they are dissimilar terms

Standard 7

Solve x: $\quad \dfrac{x}{3} + \dfrac{3x}{2} = 5$ $\times\, 6: \quad 2x + 9x = 5$ $ 11x = 5$ $ x = \dfrac{5}{11}$	
Correction: $\quad \dfrac{x}{3} + \dfrac{3x}{2} = 5$ $\times\, 6: \quad 2x + 9x = 30$ $ 11x = 30$ $ x = 2\tfrac{8}{11}$	What you do on the left, you *must* do on the right

Standard 8

Demonstrate: AE = EC	
Statement	*Motivation*
$\triangle ADE \equiv \triangle EFC$	
1. $\quad \hat{E}_1 = \hat{E}_2$	opposite angles
2. $\quad \hat{A} = \hat{C}_2$	alternating $\angle$s, DA $\parallel$ CF
3. $\quad$ AD = FC	DA $\parallel$ FC
$\therefore \triangle ADE \equiv \triangle EFC$	$\angle \angle$S
$\therefore$ AE = EC	
Correction:	
$\triangle ADE \equiv \triangle CFE$	The letters must be indicated in the *correct* order
AD = FC	AD = DB $\quad$ given
	$\quad\,$ = FC $\quad$ Opposite sides of parallelogram ... auxillary statement
	Remember! The fact that sides are $\parallel$ does not make them equal too!

Standard 9

$$\frac{27^{y+2}}{3^3} = 9^{y-1}$$	
Correction: $\dfrac{27^{y+2}}{3^3} = \dfrac{3^{3y+6}}{3^3}$ $$= 3^{3y+3}$$	• We cannot divide 3 into 27 – bases are only divided when the exponents are equal • The rule is: $\dfrac{a^m}{a^n} = a^{m-n}$ In other words the bases are made equal and only then are the exponents subtracted

$$3^{-2} = -6$$

Correction: $3^{-2} = \dfrac{3^{-2}}{1}$ $= \dfrac{1}{3^2}$ $= \dfrac{1}{9}$	You read incorrectly: You read horizontally instead of vertically! 3^{-2} is a numerator with a negative exponent. In this case the rule is: $a^{-m} = \dfrac{1}{a^m}$

Standard 10

Demonstrate that the roots of the equation are real:
$(x - k)(x - r) = p^2$

Demonstration:

$$x^2 - kx - rx + kr = p^2 \qquad\qquad a = 1$$
$$\Delta = b^2 - 4ac \qquad\qquad\qquad b = -k - r$$
$$= (-k - r)^2 - 4(1)(kr) \qquad c = kr$$
$$= k^2 + r^2 + 2kr - 4kr$$
$$= k^2 - 2kr + r^2$$
$$= (k - r)^2$$
$$\geq 0$$

Correction:

$$(x - k)(x - r) = p^2$$
$$x^2 - kx - rx + kr - p^2 = 0$$

$$\Delta = b^2 - 4ac$$
$$= (-k - r)^2 - 4(1)(kr - p^2)$$
$$= k^2 + 2kr + r^2 - 4kr + 4p^2$$
$$= k^2 - 2kr + r^2 + 4p^2$$
$$= (k - r)^2 + 4p^2$$
$$\geq 0$$

(The sum of 2 squares must be ≥ 0)

The righthand side of the equation *must* $= 0$
(Standard form)
$a = 1$
$b = -k - r$
$c = kr - p^2$

Addendum E: Fractions (Morgan, Deese and Deese)

As the computer takes such a prominent place in the lives of many students today, I specifically want to take the trouble to present the subject of "fractions" to you. In my view, the pocket calculator in particular has contributed a great deal to the deterioration in children's ability to deal with fractions – with a very detrimental effect on their mathematics in general. If you feel uncertain about any of these principles, you should immediately make sure that you catch up on your backlog.

A. Definitions

1. A fraction is one or more of the parts into which something can be divided (later on we also say that a fraction is any number in the form $\frac{h_1}{h_2}$, where h_1 and h_2 are integers).

   Examples of fractions: $\frac{1}{1}$, $\frac{1}{2}$, $\frac{2}{3}$

2. Both ways of writing fractions, namely $\frac{1}{3}$ and $\frac{1}{3}$ are correct.
3. The top part is called the numerator and the bottom part the denominator.
4. Mixed numbers such as $2\frac{1}{2}$, $2\frac{2}{3}$ consist of integers and fractions. By the way: $x = \frac{11}{3}$ is correct as an answer in a test – you need not reduce it.

B. Addition of fractions

1. $\frac{1}{3}$ and $\frac{1}{4}$ can be added up only if the denominators have been made equal – otherwise you *cannot add up*.
2. In order to make the denominators equal, you look for the *smallest* number that is a multiple of both: for 3 and 4 it is 12; for 3, 5 and 6 it is 30. Often a prime number like 7 gives problems: what you should do then, is to find the smallest multiple of the other numbers and simply multiply the answer by 7. Like this: take 4, 6, 7 and 9. The number 36 can be divided by 4, 6 and 9. Now multiply 36 with 7. $36 \times 7 = 252$. The smallest common denominator is 252.
3. Let us say you want to determine $\frac{1}{3} + \frac{1}{4}$. Now multiply the denominators with the number of times that they go into the smallest common denominator:

$$\frac{1 \times 4}{3 \times 4} + \frac{1 \times 3}{4 \times 3} = \frac{4}{12} + \frac{3}{12}$$

4. Remember that you *may (and must!)* always multiply the numerator and the denominator of any number by the same number.

5. Now you add up the numerators and place them over the smallest common denominator. Like this:

$$\frac{1}{5} + \frac{1}{11} = \frac{11}{55} + \frac{5}{55} = \frac{16}{55}$$

6. When the numerator is larger than the denominator, it is easier to first reduce it to a mixed number:

$$\frac{34}{3} = 11\tfrac{1}{3}$$

7. In algebra, it is of decisive importance that you make sure exactly what the fraction is: $\frac{2x}{3}$ is something totally different from $\frac{2}{3x}$.

C. Subtraction of fractions

1. The rules are exactly the same as for addition, except for rule 5.

2. Rule 5 now becomes: $\dfrac{1}{5} + \dfrac{1}{11} = \dfrac{11}{55} + \dfrac{5}{55} = \dfrac{16}{55}$

In other words, simply subtract the second numerator from the first one.

D. Multiplication with fractions

1. In order to multiply fractions, you have to multiply the numerators, note the answer and then multiply the denominators and note the answer, like this:

$$\frac{2}{3} \times \frac{7}{9} = \frac{14}{27}; \quad \frac{4}{5} \times \frac{6}{7} = \frac{24}{35}; \quad \frac{2}{5} \times \frac{5}{7} = \frac{10}{35} = \frac{2}{7}$$

2. In order to multiply an integer with a fraction, place the integer over 1 and then multiply. Like this:

$$3 \times \frac{2}{7} = \frac{3}{1} \times \frac{2}{7} = \frac{6}{7}$$

E. Division with fractions

1. When you divide, you turn the second number (the dividend) upside down and multiply, like this:

$$\frac{3}{5} \div \frac{2}{5} = \frac{3}{5} \times \frac{5}{2} = \frac{15}{10} = \frac{3}{2} \left(= 1\tfrac{1}{2}\right)$$

2. Remember therefore: $4 = \dfrac{4}{1}$, $x - 3 = \dfrac{(x-3)}{1}$

3. Therefore, to divide a fraction by an integer, you go about it as follows: $\dfrac{5}{9} \div 3 = \dfrac{5}{9} \div \dfrac{3}{1} = \dfrac{5}{9} \times \dfrac{1}{3} = \dfrac{5}{27}$

   You write the integer as a number over 1 and multiply by its inverse (or reciprocal).

> *And remember: when dealing with more difficult problems in algebra, you should always ask yourself: "What would this problem look like if I put it more simply, in figures?"*
> *This always works and it is your golden key to success.*

Bibliography

Allardice, B S & H P Ginsburg 1983. Children's psychological difficulties in mathematics. H P Ginsburg (ed) *The development of mathematical thinking.* New York: Academic.

Ashworth, A E 1986. *The teaching of mathematics.* Kent: Hodder and Stoughton.

Barnard, J J 1989. *Basiese wiskundige begrippe.* Paper delivered at the 13th National Convention for Mathematics, Pretoria.

Bentley, C & Malvern, D 1983. *Guides to assessment in mathematics.* London: Macmillan Education.

Bester, G 1990. Interferensie tydens die leer van wiskundige leerstof wat ooreenkomste toon. *Pedagogiekjoernaal,* Feb 1990 Vol II, no 1.

Bester, G 1988. Die verband tussen die selfkonsep van die wiskundeleerling en sy prestasie in wiskunde. *S.A. Tydskrif vir Opvoedkunde,* Aug 1988, Vol 8, no 3.

Boonstra, H H 1980. *De Rekenfout nader beschouwd.* Nijkerk: Uitgeverij Intro.

Botha, M 1990. *Slaagpunt Wiskunde.* Goodwood: Nasionale Boekdrukkery.

Carl, I M 1989. Essential mathematics for the twenty-first century. *The Education Digest.*

Christie, C 1989. *Mathematics as she is spoke in South Africa.* Paper delivered at the 13th National Convention for Mathematics, Pretoria.

Copeland, R W 1982. *Mathematics and the elementary teacher.* New York: Macmillan.

Cornelius, M 1982. *Teaching mathematics.* London: Croom Helm.

Cruikshank, D E, D L Fitzgerald & L R Jensen 1980. *Young children learning mathematics.* Boston: Allyn & Bacon.

De Jager, C, S Fitton & P Blake 1985. *Net wiskunde 9/10.* Cape Town: Maskew Miller Longman.

Du Plessis, G & W Pryba 1990. *Studievaardigheid.* Publication of the University of Pretoria, Pretoria.

Emenalo, S I & E N Okpara 1990. A study guide for mathematics students. *International Journal of Mathematical Education,* Jan–Feb 1990. Vol. 21, no 1.

Feuerstein, R 1988. *An overview of Project Intelligence.* Seminar Harvard University.

Gannon, K E & H P Ginsburg 1985. Children's learning difficulties in mathematics. *Education and Urban Society,* Aug 1985, Vol 17, no 4.

Glencross, M G & P Fridjhon 1989. An analysis of errors in high school mathematics by beginning university students. *Spectrum,* Vol 27, no 1.

Grossnickle, F E, J Reckzeh, M P Leland & N S Ganoe 1986. *Discovering meanings in elementary school mathematics.* New York: Holt, Rinehart and Winston.

Gullatt, D E 1987. How to help students in reading mathematics. *The Education Digest,* Jan 1987.

Hannan, A 1988. Should maths be multicultural? *Mathematics in school,* Vol 17, no 1.

Harper, S 1987. A child's eye view of mathematical skills across the curriculum. *Support for learning,* Vol 2, no 1.

Hudson, B 1987. Multicultural Mathematics. *Mathematics in school,* Vol 16, no 4.

Jordaan, W J, J J Jordaan & J M Niewoudt 1983. *Algemene Sielkunde: 'n Psigobiologiese benadering.* Johannesburg: McGraw-Hill.

Joubert, G J, J C Smith, P G Human & M D de Villiers 1990. Wiskundige Leesbekwaamheid. *S A Tydskrif vir Opvoedkunde,* Feb 1990, Vol 10, no 1.

Kolb, D A 1984. *Experiential Learning*. Englewood Cliffs: Prentice-Hall.

Kumon, T 1991. "It should all begin with a song". *Time*, 1 April 1991.

Lozanov, G 1978. *Suggestology and outlines of Suggestopedy*. New York: Gordon and Breach.

Marais, J L 1988. Humor in persoonlikheidsvoorligting. *SA Tydskrif vir Opvoedkunde*, Vol 8, no 3.

Maree, K 1990. *Leer jou kind lewe*. Pretoria: J L van Schaik.

Miura, I E 1987. Mathematics achievement as a function of language. *Journal of Educational Psychology*, Vol 79, no 1.

Morgan, C T, J Deese & E K Deese 1981. *How to study*. New York: McGraw-Hill Book Company.

Olivier, A 1989. Handling pupil's misconceptions in mathematics. *Pythagoras*, no 21.

Presmeg, N C 1989. Visualization in multicultural classrooms. *Focus on Learning Problems in Mathematics*, Vol II, no 1.

Resnick, L B 1983. A developmental theory of number thinking. H P Ginsburg (ed) *The development of mathematical thinking*. New York: Academic.

Schminke, C W 1978. *Teaching the child mathematics*. New York: Holt, Rinehart & Winston.

Steen, L A 1987. Points of stress in mathematics education. *The Education Digest*, Jan 1987.

Steenkamp, P du P 1981. Die voorkoms van interferensie in die foute wat senior sekondêre leerlinge in wiskunde maak. Unpublished MEd thesis, University of the Orange Free State, Bloemfontein.

Stevenson, H W & R S Newman 1986. Long-term prediction of achievement and attitudes in mathematics and reading. *Child Development*, Vol 57, no 3.

Strauss, J P 1990. Wiskunde-angs en die wiskunde-onderwyser. *Die Vrystaatse Onderwyser*, Deel 80, no 2.

Swanepoel, C H 1985. Ontoeganklikheid van wiskunde gedurende die primêre en vroeë skoolfases. *Educare*, Vol 14, no 1.

Van Rooyen, L & S Schultze 1989. *Geslagsopvoeding vir seuns*. Hammanskraal: Unibook Uitgewers.

Visser, D 1985. *Vroue en wiskunde: Fokus op geslagsverskille*. Report by the HSRC, Pretoria.

Welch, F G 1988. The continuing need for vocational education. *The Education Digest*, Dec 1988.

Witkowski, J C 1988. Solving problems by reading mathematics. *College teaching*, Vol 36, no 4, 1988.

Woodrow, D 1984. Cultural impacts on children learning mathematics. *Mathematics in school*, Vol 13, no 5, Nov 1984.

Yepsen, R B 1987. *How to boost your brain power*. Pennsylvania: Rodole Press.